Thermoplastic Aromatic Polymer Composites

a study of the structure, processing and properties of carbon fibre reinforced polyetheretherketone and related materials

Frederic Neil Cogswell

BUTTERWORTH
HEINEMANN

Butterworth-Heinemann Ltd
Linacre House, Jordan Hill, Oxford OX2 8DP

 PART OF REED INTERNATIONAL BOOKS

OXFORD   LONDON   BOSTON
MUNICH   NEW DELHI   SINGAPORE   SYDNEY
TOKYO   TORONTO   WELLINGTON

First published 1992

British Library Cataloguing in Publication Data
Cogswell, Frederic Neil
  Thermoplastic aromatic polymer composites:
  A study of the structure, processing and
  properties of carbon fibre reinforced
  polyetheretherketone and related materials.
  I. Title.
  668.9

  ISBN 0 7506 1086 7

Library of Congress Cataloguing in Publication Data
Cogswell, F. N.
  Thermoplastic aromatic polymer composites: a study of the
  structure, processing, and properties of carbon fibre reinforced
  polyetheretherketone and related materials/Frederic Neil Cogswell.
      p.   cm.
  Includes bibliographical references and index.
  ISBN 0 7506 1086 7
  1. Polymeric composites.   I. Title.
TA418.9.C6C54   1992
620.1'92–dc20   91-28715
                                                       CIP

Composition by Genesis Typesetting, Laser Quay, Rochester, Kent
Printed and bound in Great Britain by Redwood Press Limited, Melksham, Wiltshire

Contents

Foreword

At a SAMPE Conference in 1980 I lamented the lack of long fibre reinforced thermoplastic composites compared with thermosets. I said 'I believe that reinforced thermoplastics have a brighter future in the long run because the physical properties of the plastic arise from a more stringent chemistry, namely that necessary to produce linear polymerization, and hence should lead to better controlled and more reproducible physical properties'. At that time, Neil Cogswell was working on his magnificent invention which revolutionalized the field of thermo-formable composite materials by finding a commercially viable method for introducing long fibres into a high-performance thermoplastic resin.

The basic idea was very simple, as in so many inventions. The unique step was that of applying the force to the fibres in the direction in which they are strong so as to force the fibres into the resin and to coat them while preserving fibre alignment. Because the fibres are forced into the resin by applying the force along the direction in which they can sustain maximum force, they are individually wetted, the void content is kept to a minimum, and in addition the fibres can be uniformly distributed without the formation of islands of resin.

Thermoplastic composite materials of course embrace a wide field. But this book considers in detail only the material made possible by Cogswell's invention, namely carbon fibre reinforced polyetheretherketone at about 60% of volume reinforcement. Because the matrix of the resin is highly controlled, practically void-free, highly crystalline and a thermoformable plastic, many of the inherent properties of reinforced plastics and the methods of forming and manufacturing these can be demonstrated with this system. That is what this book is about and it gives me great pleasure, and indeed pride, to be able to write a foreword to it. The book covers all the difficult as well as the easy parts of composites engineering, dealing well with processing science and manufacturing technology, durability, temperature sensitivity, and environmental resistance. Neil Cogswell also recognizes that the performance of a structural material will outlast the usefulness of the structure. The issue of reclaiming high value materials is a significant one. This book deals, in an interesting way, with how thermoplastic structural materials can be re-cycled.

It is a book pointing the way to the future and I heartily recommend it to all those interested in composite materials.

Anthony Kelly
Vice-Chancellor University of Surrey

Preface

This book is designed to review our understanding of the field of thermoplastic composite materials through a detailed study of one member of that family – carbon fibre reinforced polyetheretherketone.

No prior knowledge of the field is required in order to read these chapters. The opening chapter is intended to lead the newcomer into the field. There follow detailed discussions: of the ingredients of these designed materials, of how they are made into composites, and, of their microstructure. Chapter 5 addresses the way in which such materials can be converted into structures. This is followed by a series of short reviews of the performance of carbon fibre reinforced polyetheretherketone and a discussion of the application of such materials in service. The final chapter considers the directions of research in this field and attempts to project their likely influence on the development of new businesses.

The work concludes with a number of appendices, and author and subject indexes.

Neil Cogswell

Acknowledgements

I am grateful to Mike Eades, Director of the ICI Thermoplastic Composites Business, and David Clark, Laboratory Director of the ICI Wilton Materials Research Centre, for their permission to publish this work.

In preparing the text I have received considerable assistance from my colleagues. My particular thanks are due to John Barnes who was a diligent and challenging scientific critic. The following corrected, or made material additions to, the work in draft form: David Blundell, Chris Booth, Mark Cirino, Robin Chivers, Liz Colbourn, Gerald Cuff, Mark Davies, David Groves, David Hodge, Tony Jackson, David Kemmish, Duncan Laidler, David Leach, Peter Meakin, Peter Mills, Roy Moore, David Parker, Judith Peacock, Steve Ragan, Ian Robinson, Paul Schmitz, Tony Smiley, Philip Staniland, Dave Stocks, Alan Titterton, Roger Turner, Phil Willcocks and Nabil Zahlan. I have tried to reflect their expertise, any errors that remain are my own.

My daughter Frederica read the chapters with the eyes of a scientist, but not one familiar with the field. She purged the text of jargon and pointed out where simplification or amplification were necessary.

The word processing was carried out with meticulous care and unfailing good humour by Margaret Tarry with contributions from Fiona Hack and Heather Dobson.

By her encouragement and forbearance during the long hours of writing, my wife Valerie reinforced the respect and esteem that is the basis of my love for her.

Neil Cogswell

As Materials Scientists, my colleagues and I have been encouraged, by ICI, to explore what things might be. This book is respectfully dedicated to our associates of the Engineering, Marketing and Management disciplines who actually make them happen.

1 An introduction to thermoplastic composite materials

1.1 Towards designed materials

The history of mankind can be written in terms of his use of materials. In the first place such materials were provided by nature; wood and stone are still abundant and can be both aesthetically pleasing and extremely serviceable. A second stage was to fashion those materials for a purpose: for example by hardening wood in a fire, or chipping flints to provide sharp edges. The discovery that nature also provided metals expanded the technological horizon of our ancestors and initiated a search for new and improved materials: it is no accident that, in a great many of our universities today, the disciplines of metallurgy and materials science are strongly coupled. However, in the last fifty years, following the chemical discoveries of nylon and polyethylene, the emphasis for new materials research has progressively moved towards non-metallic materials and especially the family of polymeric materials known as plastics. Simple plastics have limitations as structural materials, and it is with structural materials that man builds his world. In order to overcome these limitations recourse has been made to the expedient of reinforcing those resins with fibres to provide a family of composite materials, which, since the development of carbon fibres, have been able to compete with metals as structural materials. In this family technologists bring together the skills of diverse scientific disciplines to provide materials whose properties can be tailored to a specific need. Such structural composite materials are the heralds of the Age of Designed Materials.

Fibre reinforced polymeric composite materials have three elements: the reinforcing fibre, the matrix resin and the interface between them. Consider a piece of composite material the size of a little finger with a volume of 10 millilitres. Such a piece of composite, based on 60% by volume of carbon fibres of diameter 7 micrometers, contains 160 kilometres of fibre, 4 millilitres of resin and 3½ square metres of interface – the distance from London to Birmingham, the volume of a thimble, and the area of a large dining table respectively. The reinforcing fibres are stiff, often with a modulus higher than that of steel, but because they are also slender – 7 microns is just visible to the naked eye – they feel silky to touch. It is only when they are stuck together with resin that their rigidity becomes apparent. It is the reinforcing fibre which largely determines the stiffness and strength of the composite. For optimum effect they are highly collimated so as to provide a high volume packing: hexagonal close packing would, in theory, allow 91% by volume of fibres to be achieved, but, in practice, most commercial structural composite materials utilize 60–65% by volume of fibres.

High collimation of the fibres provides the maximum possible reinforcing effect along the axis of the fibre. Transverse to that direction there is little effective reinforcement. This anisotropy of properties provides added freedom to the designer, who can now arrange the reinforcement in his materials according to the load paths in the structure that he is building. The matrix resin transfers stress between the fibres and stabilizes those fibres when the structure is subjected to compression loadings. It must provide this service when under attack by hostile environments. The resin is also the medium which determines the processes by which the composite material is shaped into a structure and, during that process, it protects the reinforcing fibres from damage by attrition. The stiffness and strength of the resin is usually low in comparison to that of the fibres – a factor of 100 between reinforcement and resin modulus is common – but the resin plays a vital role both in determining the serviceability and the processability of the material.

The third element in a composite material is the interface between the resin and the reinforcement. The physical constraints provided by a high loading of fine diameter fibres is often sufficient to affect the properties of the resin in the neighbourhood of the fibre. In addition the fibres themselves may have chemically activated surfaces or be coated to promote adhesion of the resin. Thus, instead of a simple interface between materials, it may be more correct to consider an interphase. In a system containing 60% by volume of reinforcing fibres 7 μm in diameter the mean thickness of resin coating on each fibre is 1 μm and any 'interphase' may extend through the whole of the matrix. The integrity of the interface plays a critical role in optimizing the service performance of the composite material, especially in respect of the resistance to hostile environments and in determining the toughness of a composite. The reinforcement, matrix and interface act together to achieve optimized serviceability.

There are several texts[1-10] that describe the science of fibre reinforced composite materials – their history, application and potential – as viewed from different perspectives. Particular emphasis is naturally placed upon their mechanical performance, which depends upon their laminar construction. The basic building block of composite structures is a simple lamina of collimated fibres typically ⅛ mm thick (Figure 1.1).

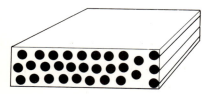

Figure 1.1 Basic building block: a lamina of collimated fibres

Along the fibre direction the theoretical stiffness of the composite lamina is given by:

AXIAL STIFFNESS $= V_F E_F + V_M E_M$

where V_F is the volume fraction of the reinforcing fibre
V_M is the volume fraction of the matrix resin
E_F is the modulus of the reinforcing fibre
E_M is the modulus of the matrix resin

Usually the modulus of the fibre is much greater than that of the resin, and a good approximation to the axial stiffness of the composite can be made by ignoring the resin component. Transverse to the fibre direction there is much less reinforcement. In this direction the fibres themselves may be significantly less stiff than along their axis, and since their cross-section has a very low aspect ratio, they will have little reinforcing effect but simply act as a filler. As a first approximation the transverse stiffness can be taken as:

TRANSVERSE STIFFNESS $= E_M / V_M$

In general, composite materials are not used as simple axial laminates, because structures have to accept a variety of loading patterns. To accommodate such loads, simple shell structures are usually laminated with the individual plies or laminae oriented in different directions (a description of such laminates is given in Appendix 1). A quasi-isotropic laminate is one in which the composite layers have been arranged so that, in the plane of the moulding, the properties are the same in any direction. High stiffness and strength are only part of the equation in the design of structures. Compared to metals, composite materials have a relatively low density: the single outstanding property of carbon fibre reinforced plastics is their stiffness per unit weight (Table 1.1).

Table 1.1 Comparative weights of panels with same bending stiffness

Material	Modulus GN/m^2	Relative thickness	Specific gravity	Relative weight
Metal				
Steel	210	1	7.8	7.8
Aluminium	73	1.4	2.7	3.8
Aircraft grade composite				
Quasi-isotropic	40	1.7	1.6	2.7
Uniaxial	120	1.2	1.6	1.9
Space grade composite				
Uniaxial	300	0.9	1.6	1.4

The low density of composite materials increases the thickness of a panel. This further increases the stiffness, allowing composites to achieve 30% weight-saving in comparison to aluminium structures of the same flexural stiffness. There are still the through thickness properties of the laminate to be considered. In general, structures are not designed to carry loads in this direction but adventitious loads may occur, and the design of composite structures should also take account of those through thickness properties. This is one of the areas where we shall see that thermoplastic composites have demonstrated exceptional advantage.

A further major structural advantage of composite materials is that they have proved themselves to be exceptionally resistant to fatigue in comparison with metals. In addition, because it is possible to design the material at the point of manufacture, it is possible to tailor in certain features, such as dielectric properties, to achieve highly specific effects: one of the earliest and most successful uses of glass fibre reinforced plastics has been in the provision of radomes in aircraft[11]. A further service advantage is resistance to corrosion, where polymeric materials have an excellent record.

As well as their advantages in service, the introduction of polymeric composite materials can significantly reduce the total part count in building a structure. This leads to economies of manufacture. There is one area where composite materials have yet to attain their full potential. At this time the field of structural design is, in practice, largely based on a metals history, i.e. design practice is to assume isotropic materials. At the present time many of the structures that are built from composite materials are based upon modified 'metal' designs, but there are notable exceptions where the principle of anisotropy is a cornerstone. The theory of anisotropic design is established[12] but it has only become so during the last twenty years. It will be of great significance to the future of designed materials to see how those who have been schooled in anisotropic design apply those principles in practice.

As well as the bulk properties of materials, there is a growing opportunity to tailor the surface of a material. At this time such tailoring in composite materials tends to be at the interface between fibre and resin or a surface modification to functionalize two surfaces in order to facilitate joining them together. Ultimately that principle of surface modification will also be expressed as a way of adding service function to the exterior of a structure.

Our present generation of reinforced plastics is only making relatively limited use of the composite concept. Fibres can carry messages as well as loads; the incorporation of piezo-electric elements in a structure can make it responsive; the notion of combining 1 kg of computer for every 10 kg of structure would provide memory and adaptive learning ability; the principle of 'smart' materials is, of necessity, a composite concept. The present generation of composite materials offers significant advantages to the designer of today; the designer of tomorrow has the opportunity to tailor those materials to the needs of the function he seeks to achieve.

From this introduction it is necessary to refine the objectives of this work. It is my belief that, provided the broad picture is retained in view, the study in depth of a narrow part of the field of the science of materials can lead to the proposition of general principles whose validity can subsequently be tested. Thus I seek to provide that study in depth within a narrow field. This study provides a self-consistent body of data from which the relation between different functions of the material can be established. The organization of this work, which necessarily encompasses a wide range of scientific disciplines, should then provide a picture against which other systems may be compared and contrasted. Lastly, it is desirable that the subject matter of the detailed study should be both timely and of lasting value. These considerations lead to the selection of thermoplastic structural composites and, in particular, their paradigm – carbon fibre reinforced polyetheretherketone – as the theme.

1.2 Why thermoplastic structural composites?

In 1990 thermoplastic composites represented about 3% of the total market for polymer matrix structural composites. The greater part of organic matrix composites presently in service are based on crosslinkable thermosetting resins. Of these, epoxy resins are the most prolific representatives. When carbon fibres were first introduced, there were several attempts[13–17] to develop composite materials based on thermoplastic polymers such as the polysulphone family. By creating low viscosity polymer solutions, the reinforcing fibres could be impregnated with such resins[18]. However, it proved extremely difficult to eliminate residual solvents from such materials[19, 20], and their retention caused problems, including blistering of the moulding and a reduction in properties, particularly at high temperature. Further, the use of solution technology betrayed a sensitivity of those amorphous polymers to attack by chemical reagents. Since composites were of particular interest to the aerospace industry, where hydraulic fluids, aviation fuels and the use of paint strippers are widely encountered, soluble materials were placed at a severe disadvantage. Not all thermoplastic composites were made by solution processing: it was possible to interleave layers of reinforcing fibres and resin films and then, in a protracted moulding operation, wet out the fibres in a high-pressure moulding process[21–3]. Such film stacked composites gained a small niche in the market place but, for the most part, the field of structural composite materials was ceded to ccrosslinkable thermosetting resins. Because of the convenience with which epoxy resin systems could be hand layed up into structures, thermosetting composite materials became established as excellent prototyping materials and were gradually translated into full scale production.

By 1980, three problems had become evident with epoxy composite materials: their brittleness, their sensitivity to water, and the slow manufacturing technology. Although it was possible to improve either the toughness or the water resistance of epoxy composite materials, a common experience was that an improvement in one property usually led to a falling off in the other. The introduction of preimpregnated carbon fibre reinforced polyetheretherketone, a high temperature semi-crystalline polymer, in continuous tape form[24] simultaneously offered a solution to the problems of toughness and environmental resistance, with the added bonus of potentially developing high rate processing: thermoplastic materials were translated into a major platform for research and product development. These material forms are now becoming qualified products. They extend the range of applications established for their thermosetting cousins in particular into that segment of industrial activity that requires high rate automated fabrication.

Thermoplastic matrix materials derive their properties from their long chain entangled molecules. The entanglement of the chains provides the matrix with strength and effectively acts as a temporary crosslink. Because the chains are not fixed by their chemical structure, they have the ability to slip past one another when subjected to intense local stress; thus, whereas a crosslinked polymer has no alternative but to be strong or to break in a brittle fashion, a linear chain polymer has the ability to dissipate energy locally by chain slippage. This ability confers the property of toughness on the composite. Some linear chain polymers also have the ability to pack tightly together in a semi-crystalline network. This network also provides an effective physical crosslink in the system, enhancing strength. The crystalline structure further provides resistance to attack by hostile reagents. When a linear chain, thermoplastic, polymer is melted the chains may move freely with respect to one another, so that the composite can be formed into shape. In contrast to their thermosetting cousins, with long chain thermoplastic materials this shaping process is a purely physical one, depending on heat transfer, geometry and the forces applied. The absence of chemistry, the necessary companion of thermosetting processing, means that thermoplastic polymers can be fabricated into structures rapidly and with a high level of quality assurance: all the chemistry has been carried out by those people whose business is chemistry, leaving the fabricator free to concentrate wholly upon his art of making shapes.

The potential for rapid processing with thermoplastic structural composites is of particular significance. With high performance thermosetting resins the cure cycle is usually several hours long: with thermoplastics the absence of chemistry means that structures can be formed in minutes, or even seconds. The protracted nature of thermoset processing is not necessarily a constraint in industries, such as aerospace, where the required rate of production may be less than one a day and each individual structural component must be optimally tailored to shape. In such circumstances the advantages of low capital expenditure on tooling may justify the use of slow, labour intensive, fabrication processes. Such a vision, however, unnecessarily limits the horizon of structural composite materials. For the first twenty years of their life it was widely believed that carbon fibre reinforced plastics would make the transition from specialist aerospace materials to general industrial products: considerable investment in carbon fibre capacity testifies to that belief, but, except in the area of high performance sporting goods and short fibre moulding compounds, the penetration has been disappointingly small. The difficulty in achieving that translation has been the requirement for appropriate mass production technology for high performance structural composites. Thermoplastic structural composites offer the potential for mass production, which may provide the key to the metamorphosis of high performance composite materials into high quality mass production materials for the general industrial designer.

The thermoplastic family of materials is particularly broad in respect of the service it can offer. There are a wide range of resins: elastomeric materials such as polyurethanes, general purpose resins such as polypropylene, engineering plastics such as nylon and polycarbonate, and high performance polymers such as polysulphones and polyketones. Plastics can be derived from each of these polymers in a wide range of forms, including:

neat resin for extrusion or moulding;
fibre, film and foam forms;
filled compounds;
short fibre reinforced injection moulding materials;
long fibre moulding systems;
random fibre mat stampable sheet systems;
and continuous fibre structural composites;

In this family the reinforcement aspect ratio and content varies (Table 1.2).

Table 1.2 Reinforcement aspect ratio for thermoplastic product forms

	Volume fraction filler (%)	Fibre length in moulding	Reinforcement aspect ratio
Structural composite	60	Continuous	100 000
Stampable sheet	30	25 mm	2 000
Long fibre moulding compounds	30	2.5 mm	200
Short fibre materials	20	0.25 mm	20
Filled compounds	10–40	particulate	2

This family provides the environmentally friendly option of reclaiming offcuts from structural composite materials, or indeed lifetime expired mouldings, into lower performance moulding compounds by the addition of extra resin and reducing the fibre length mechanically. Further, there is the option of using high performance thermoplastic composites as an integrated structural component in a moulding based on discontinuous fibre or unreinforced resin. Given the ability sequentially to process thermoplastic composites, the potential for innovative integrated design is broad indeed.

Thermoplastic structural composites are relative newcomers to the family of designed materials. Nevertheless a large body of published literature is already available[25]. Some five hundred scientific papers have already been contributed to the open literature describing the thermoplastic composite paradigm carbon fibre reinforced polyetheretherketone, and it is possible that, in the open literature, we actually know more about this material than any other single composite system. What is learnt on one member of the family of linear chain polymers, particularly about the arts of processing and component design, can be readily translated for use with other members of the same genus. What we see in the open literature is only the tip of the knowledge iceberg. Each year the United States Airforce holds workshops reviewing the experience of these materials, and several 'closed' conference sessions discuss these materials in America. The time is opportune to review our knowledge and understanding of these materials in order to provide a foundation upon which the product designer can build his experience with confidence and from which the materials designer can evolve enhanced materials systems to meet future challenges.

There are several review articles[26–38] that chart the significance, potential and growth of the field of thermoplastic composites. Not surprisingly, many of the papers emanate from the raw material supplier side of our industry but there is also a prestigious review[39], prepared for the National Materials Advisory Board of the US National Research Council, by a panel of academics and representatives of the aerospace user industry that provides an independent consensus.

1.3 Composite science

Besides the specific elements of fibre science, polymer chemistry and interface technology from which our composite materials are derived, it will be necessary to integrate a range of disciplines. How those elements are assembled will also be important: morphology, local order and orientation will be critical factors. To be useful, the products must be shaped; thus their rheology is of consequence. Once shaped, their service performance must be understood, in particular their mechanical performance and how that behaviour is influenced by environment. The behaviour of laboratory specimens is only a part of the story: the behaviour of real structures depends also on their design, and so we must understand the science of that step too. All these elements will interact: thus, the properties of a given structure may depend, not only on the shape, but also on the local morphology of the material. That morphology can be influenced by the basic composition of the material and by how it has been processed. The selection of processing technology will depend upon the design of the component and upon economic constraints. This text seeks to define the individual elements of this composite science and to describe how those interactions occur. Ultimately we must seek a fully integrated composite science so that we can design material, structure and product function simultaneously.

In considering polymeric composite materials we must be aware of effects at a range of scales. (Figure 1.2). At the smallest, nanometre, scale we must consider the architectural substructures of the polymer matrix resin molecule and that of the reinforcing fibre, together with their critical interaction at the interface between matrix and reinforcement. The interfibre spacing, which is occupied by the resin, is of the order 0.1 to 10 µm. The reinforcing fibres themselves are usually about 10 µm in diameter, and this size scale often describes the spherulitic texture of a semi-crystalline polymer. The primary thickness scale of the laminate structure, within which layer the fibres are highly collimated, is about 0.1 mm, and this scale may also be encountered in the bondlines where composite components are assembled into composite structures. The thickness scale of composite components is usually in the range 1–100 mm and their most frequent length scale is in the range 0.1 to 10 m. Assembled structures are usually in the 1 to 10 m range, but can, of course, be considerably longer. Within a given component, fibres will usually

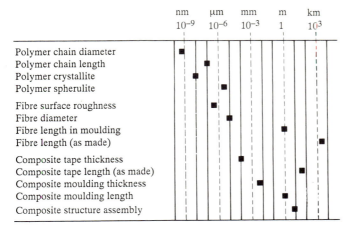

Figure 1.2 Scales of polymeric composite materials

extend the full length of that structure, but they may, for example in a filament wound structure, be very much longer. The manufactured, cut to length, scale of the preimpregnated tapes, which constitute the major building blocks for composite structures, ranges from 1 m to 1 km or more in length, while the fibre tows from which they are prepared are typically 1 to 10 km long. Thus, in composite science, we must be aware of effects from the nanometre to kilometre scale – twelve orders of magnitude. This book will be seeking to define effects at each of these scales and how they influence performance in the final structure.

Each of the disciplines which comprise composite science are expert fields in their own right. The only other book which specifically addresses this field[40] is a compilation of expert texts, and it is to the expert that we must defer for discussion of detail. This text seeks to present an overview and integration of such expert views interpreted from the standpoint of the general scientist, although I must declare a particular interest in the technologies of producing such composite materials and of manufacturing artefacts from them.

References

1–1   J. E. GORDON, *The New Science of Strong Materials*, Penguin Books (1976).
1–2   E. FITZER, *Carbon Fibres and Their Composites*, Springer Verlag (1983).
1–3   J. A. PEACOCK AND F. N. COGSWELL, 'Fibre Reinforced Advanced Structural Composites', in *Multicomponent Polymer Systems*, edited by I. Miles and S. Rostami, Longmans (in press).
1–4   I. K. PARTRIDGE, *Advanced Composites*, Elsevier Applied Science (1990).
1–5   J. C. HALPIN, *Primer on Composite Materials*, Technomic Publishing Co (1980).
1–6   D. HULL, *An Introduction to Composite Materials*, Cambridge UP (1981).
1–7   A. KELLY AND N. H. MACMILLAN, *Strong Solids*, 3rd edition, Clarendon Press, Oxford (1986).
1–8   A. J. M. SPENCER, *Deformations of Fibre-Reinforced Materials*, Clarendon Press, Oxford (1972).
1–9   R. M. JONES, *Mechanics of Composite Materials*, McGraw-Hill (1975).
1–10 T. W. TSAI, *Composites Design*, 4th edition, Think Composites, Dayton, Ohio (1988).
1–11 See 1–1.

1–12 See 1–10.

1–13 J. McAINSH, 'The reinforcement of polysulphones and other thermoplastics with continuous carbon fibre', BPF 8th International Reinforced Plastics Conference, (1972).

1–14 D. J. LIND AND V. J. COFFEY, 'A method of manufacturing composite material', British Patent 1 485 586 (1977).

1–15 L. N. PHILLIPS, 'The properties of carbon fibre reinforced thermoplastics moulded by the fibre stacking method', RAE Technical Report 76140; *Fabrication of Reinforced Thermoplastics by Means of the Film Stacking Technique,* HMSO, London, 1980.

1–16 P. A. HOGAN, 'The production and uses of film stacked composites for the aerospace industry', SAMPE Conference (1980).

1–17 J. T. HOGGATT, 'Thermoplastic Resin Composites', 20th National SAMPE Conference (1975).

1–18 See 1–13.

1–19 N. J. JOHNSTON, T. K. O'BRIEN, D. H. MORRIS AND R. A. SIMONDS 'Interlaminar Fracture Toughness of Composites II: Refinement of the Edge Delamination Test and Application to Thermoplastics', 20th National SAMPE Symposium, pp. 502–517 (1983).

1–20 K. E. GOODMAN AND A. R. LOOS, 'Thermoplastic Prepreg Manufacture', *Proc. Am. Soc. for Composite Materials,* 4th Tech Conf., 746–754 Technomic (1989).

1–21 See 1–14.

1–22 See 1–115.

1–23 See 1–116.

1–24 G. R. BELBIN, I. BREWSTER, F. N. COGSWELL, D. J. HEZZELL AND M. S. SWERDLOW, 'Carbon Fibre Reinforced PEEK: A Thermoplastic Composite for Aerospace Applications', European Chapter of SAMPE Meeting at Stresa (1982).

1–25 F. N. COGSWELL, 'Thermoplastic Structural Composites: Evolution and Revolution', in 'Materials and Processing – Move into the 90's', *Proc. 10th Int. Eur. Chapter Conf. of SAMPE,* Materials Science Monographs **55**, pp. 1–10, Elsevier (1989).

1–26 *Ibid.*

1–27 F. N. COGSWELL AND D. C. LEACH, 'Continuous Fibre Reinforced Thermoplastics: A Change in the Rules for Composite Technology', *Plastics and Rubber Processing and Applications* **4**, 271–276 (1984).

1–28 P. E. McMAHON, 'Thermoplastic Carbon Fibre Composites' printed in *Development in Reinforced Plastics – 4,* edited by G Pritchard, pp. 1–30 (1984).

1–29 S. CHRISTENSEN AND L. P. CLARK, 'Thermoplastic Composites for Structural Applications: An Emerging Technology', 31st International SAMPE Symposium, pp. 1747–1755 (1986).

1–30 N. J. JOHNSTON AND P. M. HERGENROTHER, 'High Performance Thermoplastics: A Review of Neat Resin and Composite Properties', 32nd International SAMPE Symposium, 6–9 April 1987.

1–31 G. R. BELBIN AND P. A. STANILAND, 'Advanced Thermoplastics and their Composites', *Phil. Trans. R. Soc. Lond.,* A 322, pp. 451–464 (1987).

1–32 F. N. COGSWELL, 'Thermoplastic Structural Composites', Institute of Physics Conference, Serial No. 89, *'New Materials and their Applications',* Warwick 1987, pp. 77–87, IOP Publishing Limited (1988).

1–33 National Materials Advisory Board, *'The Place for Thermoplastic Composites in Structural Components,* National Research Council NMAB-434 (1987).

1–34 I. Y. CHANG AND J. K. LEES, 'Recent Development in Thermoplastic Composites: A Review of Matrix Systems and Processing Methods', *J. Thermoplastic Composite Materials,* **1**, pp. 277–296 (1988).

1–35 G. R. GRIFFITHS, 'Thermoplastic Composites, Past, Present and Future', in 'Materials and Processing – Move into the 90's', *Proc. 10th Int. Eur. Chapter Conf. of SAMPE',* Materials Science Monographs **55**, pp. 101–110, Elsevier (1989).

1–36 R. HASLER, 'Continuous Fibre Reinforced Thermoplastic Composite Material', Conference on High Performance Materials 'Hipermat 89', London (1990).

1–37 G GRAY AND G. M. SAVAGE, 'Advanced Thermoplastic Composite Materials', *Metals and Materials,* **5**, pp. 513–517 (1989).

1–38 D. C. LEACH, 'Continuous Fibre Reinforced Thermoplastic Matrix Composites', in *Advanced Composites,* edited by I. K. Partridge, Elsevier Applied Science Publishers (1989).

1–39 National Materials Advisory Board, *op. cit.*

1–40 L. A. CARLSSON, *Thermoplastic Composite Materials,* Elsevier (1991).

2 Components of a thermoplastic structural composite

In order to understand the serviceability of a composite system it is first necessary to consider its ingredients. For a structural composite material those ingredients are the reinforcing fibre and the resin, and the interface, or interphase, between them. In this case we define the system by the thermoplastic nature of the resin, which we must explore in some depth. There are important differences between classes of thermoplastic resins from which special properties of the composite derive. We shall pay particular attention to the semi-crystalline polymer polyetheretherketone. The fibres employed are those used in a wide range of composite materials and have received a very adequate review in other works[1,2]; nevertheless it is appropriate to summarize their basic features and, in particular, define the properties of high strength carbon fibres, which provide the backbone of the industry. The definition of fibre concentration and resin – for example 60% by volume carbon fibre in polyetheretherketone does not totally describe the composite. In this chapter we must also consider those features that influence the adhesion between the fibre and the resin – the interface or interphase region. The quality of a composite material, and, in particular, the quality of the interface region, depend upon the way in which the components are assembled: this stage will be considered in Chapter 3. While all thermoplastic structural composites share certain features in common because of similar ingredients, it is necessary to optimize the integration of those ingredients to achieve the best possible product.

2.1 Thermoplastic matrix resins

Four families of thermoplastic resin can be considered as potential matrix resins for composites: linear chain extendable polymers, fully polymerized amorphous polymers, liquid crystalline polymers and fully polymerized semi-crystalline polymers. In the family of high performance resins all share a common chemical theme in the use of rigid ring structure elements in the chain backbone. It was from the extensive usage of aromatic rings in the backbone of the 'Victrex' range of polymers from ICI that composites based on that family derived their name Aromatic Polymer Composite. This became abbreviated to the acronym APC, which, in turn, has also come to stand for advanced polymer composite. There are three key properties of thermoplastic structural composites: toughness, environmental resistance and processability. All thermoplastic resins can provide toughness but they differ significantly with respect to their environmental resistance and processability.

The properties of the matrix resin are critical to each stage in the life of the composite. At the stage of manufacturing the preimpregnated product form, there is a requirement for low viscosity to wet out the fibre surface adequately. For conventional impregnation technologies, using thermosetting resins, viscosities significantly less than $1 \, Ns/m^2$ are preferred. Such resins are pourable and will usually wet out fibres as a result of surface tension forces. With thermoplastic polymers, such low viscosities are not available in the melt phase, and special impregnation techniques must be employed to wet out the fibre with melt, or recourse must be made to solution, emulsion, dispersion or powder processes to coat the surface of the fibre with the resin before fusing it into place. Viscosity also dominates the processing stage where the preimpregnated product, or 'prepreg', is formed into its desired shape: too low a viscosity leads to excess resin migration; too high a viscosity unduly constrains the consolidation of the laminae and the relative movement of the fibres required in some shaping processes. There is a general consensus that the preferred melt viscosity range for most forming processes is about 10^2 to $10^3 \, Ns/m^2$, but very much higher viscosities can be accommodated when using slow processing methods. Besides a preference for a melt viscosity in this range the temperature under which the material can be formed is of particular importance. Today there are a wide range of technologies that can operate at temperatures up to 400°C, but working at higher temperatures is acknowledged to be particularly difficult. Besides the temperature, a broad processing window, within which the mouldings will have consistent properties, is especially desirable. In this context ±20°C from the optimum temperature is considered satisfactory. At the processing temperature the polymer should be thermally stable for at least 1 hour: while thermoplastics are usually intended for rapid forming, some very large structures can only be made in relatively slow operations, owing to the time taken to heat up and cool down the structure. Another aspect of the processing window is that the morphology of the resin, and so its properties, should be independent of such factors as cooling rates within the range 5°C/min to 500°C/min usually encountered in practice. For service performance the resin should be stiff and strong. Usually this is reflected by a preferred tensile modulus of at least 3 GPa and tensile strength of at least 70 MPa. However, it is actually the shear properties of the resin which are most important (Appendix 2). A particular feature of thermoplastic composites is their outstanding toughness. High resin toughness, combined with good adhesion between matrix and fibre, is greatly prized[3]. Some evidence suggests that, for the major manifestation of toughness – damage tolerance – quite modest values of fracture toughness, are adequate[4]. Other experience indicates that, if truly outstanding properties in such areas as resistance to transverse ply cracking and good tribological properties are to be achieved, higher levels of toughness are preferred. The resin must also provide for protection against attack by hostile environments, ranging from water, through solvents to fire. Not all these factors can easily be met simultaneously: high stiffness and toughness speak of high molecular weight, rigid chains; low viscosity is most commonly found in low molecular weight, flexible systems; good resistance to solvents is usually found with semi-crystalline polymers but broad processing windows are most easily achieved in amorphous materials. In designing a resin system for composites some optimization must usually be sought, dependent on the function to be served.

2.1.1 Chain extendable resins

This class of resins, which presently finds its chief expression in the polyimide family[5,6], are really prepolymer systems, where the polymer is actually formed in a thermosetting reaction once the structure has been made. This approach has the obvious advantage that, since the prepreg stage uses small molecule polymer precursors, conventional methods may be used to wet out the fibres to give product forms that retain the tack and drape of the well known epoxy resins. This enables the standard technologies of hand lay up and autoclave cure to be exploited. One drawback of this family is that the chain extension usually occurs by a condensation route and requires a large volume of volatiles to be extracted during polymerization. Unlike their crosslinking cousins, which have multi-functionality, these resins are constrained to polymerizing from the ends of the chain only. It follows that, in the final stages of polymerization, the chains are highly restricted by their neighbours. It can be difficult to mop up the final unreacted prepolymer: very protracted cure cycles may be required. The processing of composites based on such resins has been considered in detail in the publications of Gibbs[7]. Once fully polymerized, these materials are reputed to have some thermoplasticity. Faults in the laminates can be healed by a high temperature moulding process, but no evidence of thermoplastic shaping has been reported. It is at least possible that the chain extension reaction includes some element of crosslinking that immobilizes the chain; certainly moulding resins of this family[8] have exceptionally high melt viscosities. The polyimide family of resins generally has excellent resistance to solvents[9] and temperature, although some members do not perform well in hot, wet conditions. Of all the isotropic linear chain polymers, the polyimide resins have the highest stiffness: this property helps to stabilize the fibres under compression loading. These polymers, because of their linear chain structure, have the ability of internal energy dissipation by entanglement slippage, leading to enhanced damage tolerance, an outstanding property of the thermoplastic family. They are therefore sometimes defined as members of that family. In respect of their processing – the definitive stage of a thermoplastic – the potential for high rate forming has not been demonstrated.

Impregnation from monomer and subsequent polymerization has also been exploited in the case of acrylic polymers[10] and styrenes[11]. In some cases additional polymer is dissolved in the monomer, and that polymer can subsequently form a blend phase. The polyetherimide family of resins can also be polymerized in a melt reaction process[12]. The chemistry that allows this imidization to take place in the melt can also be exploited in the formation of the polymer *in-situ* after impregnation with low viscosity prepolymer. One variant on the chain extension theme which leads to genuine thermoplastic composites is the use of low molecular weight polymers to form prepregs. The subsequent reaction of those chains together forms high molecular weight systems[13]. In particular, in the case of composite materials, the activating catalyst can be applied to the reinforcing fibres[14] and in this way especially tough interfaces can be formed. This method has been proposed for use with ring-opening polymers[15]. Colquhoun[16] has demonstrated a ring-opening polymerization route to polyetherketones. In contrast to condensation reactions, large volumes of volatile material are not formed, but a drawback of the catalysed system is that it can be difficult to terminate the reaction. The number of times such a material can be reprocessed without significantly altering the structure is limited. Achieving a well defined chemical structure requires particular attention to the chain extension processes.

In some thermoplastic polymers, for example polyarylene sulphides, chain extension happens naturally at the processing temperature. As well as simple chain extension leading to enhanced toughness, some branching or crosslinking can occur. Those processes reduce crystallinity[17] and cause the material to become intractable. This characteristic necessarily limits the extent to which such polymers can be reprocessed.

2.1.2 Amorphous thermoplastics

The first thermoplastic polymers to be considered as matrices for structural composites were the amorphous polysulphone family[18]. The term amorphous implies that the polymer chain is present in a random coil without any high degree of local order, as would be present in a semi-crystalline polymer. The main advantage of amorphous polymers, and their principal drawback, is that they can usually be easily dissolved in a range of convenient industrial solvents. The advantage is that this means they can be prepregged by conventional low viscosity means. The disadvantage is that the ability to make solutions betrays the potential for attack by such solvents in service. This sensitivity to solvent attack initially relegated such materials to non-structural applications such as aircraft luggage bay liners, where their good fire, smoke and toxicity characteristics combined with toughness could still be exploited. Another problem associated with solvents is the difficulty of eliminating all the residual solvents after prepregging (Chapter 3). The presence of residual solvent is widely recognized as a major problem, resulting in reduction of glass transition temperature and defects in the composite moulding[19,20]. It also causes concern for the environment. Of course the residual solvent can actively be exploited to provide a tacky prepreg to facilitate hand lay up technologies: the residual solvent is subsequently removed as part of the shaping processing operation. Amorphous polymers can also be prepregged from the melt[21], thereby eliminating the concern about residual solvents; however, this is generally conceded to be a more difficult operation. One major advantage of this family is that, for a given melt processing temperature, a higher glass transition temperature, Tg, is usually available with amorphous materials than with their semi-crystalline cousins. As a rule of thumb[22], the usual processing temperature for a semi-crystalline polymer is Tg +200°C, whereas in the case of amorphous polymers the temperature is Tg +100°C or even less. Since all organic polymers appear to have a limitation on processing temperature of about 420°C, because of thermal decomposition, there is a clear potential for amorphous materials to supply the higher temperature matrix resins. Note, however, this limitation is not a law of polymer science, only an empirical result of common experience, which, in time, polymer chemists may be able to overcome. One factor that has been asserted in favour of amorphous systems is that they do not crystallize, so that there is one less 'variable' to consider. This is true, but those who cite that advantage are usually silent on the issue of free volume annealing, which gradually changes the properties of amorphous materials with time. This change can be particularly noticeable in the case of ageing at temperatures just below Tg. In the case of semi-crystalline polymers these amorphous regions are less important and such ageing is less evident. A real advantage for amorphous polymers is that there is a lower change in volume on solidification from the melt. Since there is no step change in density associated with the formation of crystalline regions, such materials are less subject to distortion on cooling from the processing operation and, in the case of composite materials, lower levels of internal stress may be generated. However, the majority of such internal stresses build up between the glass transition temperature of the polymer and ambient; consequently, if high Tg amorphous polymers are being used, that advantage may be masked. Amorphous polymers also give a cosmetically satisfying glossy surface finish. Obviating this advantage, amorphous polymers tend to be more subject to creep and fatigue than semi-crystalline polymers. These all tend to be secondary issues besides the critical question of environmental resistance, which remains the major stumbling block for this class of materials. Not all structural applications require the outstanding environmental resistance of semi-crystalline polymers. The aerospace industry is looking once more at the amorphous thermoplastic polymers and finding significant application areas for them, especially where high temperature performance is required and solvent susceptibility can be accommodated.

2.1.3 Orientable polymer matrices

A technologically important feature of linear chain polymers is that the molecules can be oriented, usually by mechanical work, to provide enhanced stiffness and strength in a preferred direction. This facility is exploited most obviously in the manufacture of fibres and films. It is also frequently found as a significant factor in thermoplastics made by extrusion or injection moulding, particularly where the plastic is being shaped and solidified simultaneously, so that the long chain molecules have insufficient time to forget their deformation history. In thermoplastic polymer composites the level of local deformation during melt processing is small and the stress relaxation characteristic of the melt is sufficiently rapid that the molecules relax to their random coil configuration: there is usually little orientation of the polymer matrix resulting from conventional fabrication processes.

There is one class of thermoplastic polymer – often called liquid crystal or self reinforcing polymers – wherein the molecules are very readily oriented in the melt, and that orientation persists long after the stress which caused it has been removed. The most widely explored family of liquid crystal polymers is that of the aromatic polyesters[23]. Liquid crystal polymers as a family have several properties that recommend them as matrices for composites. In respect of service performance they have high tensile stiffness and strength parallel to the orientation, high service temperatures, low thermal expansion and excellent resistance to chemical reagents. Because of their rod like structure there is little or no entanglement between neighbouring molecules: their melts have surprisingly low viscosities which should facilitate impregnation of the fibre bundles. Added to these advantages is the design vision of being able to control the orientation of the matrix phase independently of that of the reinforcing fibres: for example, the weakness of composite materials transverse to the orientation of the reinforcing fibre could be compensated for by orientation in the matrix phase. This is a very real catalogue of potential advantage, which is offset by one major weakness. The primary mechanical duties of the matrix phase is to redistribute stress from one fibre to its neighbours and to support the fibres when they are in compression. At the micromechanical level these duties are performed largely through shear modes of deformation (see Appendix 2). In shear, highly oriented polymers do not perform well. Indeed, in the case of injection moulding plastics, significant performance advantage can be gained by introducing short reinforcing fibres in order to disrupt and entangle the highly oriented domains in liquid crystal polymers, thereby reducing their anisotropy[24]. Thus, for the general family of structural composite materials, high molecular orientation of the matrix is not usually desirable; however, I am certain that the future will see the use of this important family of resins in composite materials, and this will be most evident where function besides mechanical performance are to be addressed.

The concept of using rodlike polymer molecules in conjunction with random coil polymers to provide a 'molecular composite' has also been extensively explored[25]. The driving force behind this research, largely sponsored by the United States Air Force, has been to provide a family of thermoplastic structural composites intermediate between the liquid crystalline polymers and conventional fibre reinforced composites. If the principle were extended to its extreme of single molecule separation, the weaknesses of thermoplastic liquid crystal polymers in respect to stiffness and shear properties could be overcome. In a molecular composite the rodlike polymer which acts as reinforcement would not be required to melt and the random coil matrix molecule would supply the required shear translation between the rods. Because such a material would be an all polymer composite, advantages in lower density and electrical properties could be envisaged in comparison with carbon fibre reinforced materials. By careful selection of the chemical structure of the two polymer phases, compatibility at the interface might be achieved. Further, the molecular scale of the material would mean that larger scale heterogeneities were avoided, giving enhanced reliability and a potential to design miniaturized composite systems. In respect of mechanical properties the rod like molecules would be equivalent to high aspect ratio fibres, allowing good property translation, but nevertheless sufficiently short in absolute length to allow the composite to be processed by high rate thermoplastic technologies such as injection moulding. These are all powerful and pressing physical advantages, but they have encountered an equally pressing and pragmatic thermodynamic problem. In general, rods and coils do not mix (Appendix 3). Tenacious research towards molecular composites has explored many routes to overcome this difficulty. At the present time I am not aware of molecular composites containing a significantly high volume fraction of rod like molecules where individual molecule separation has been demonstrated: the results available suggest bundles of the order of ten rod like molecules where, inevitably, only a small fraction of their total surface is wetted by the random coil phase.

2.1.4 Semi-crystalline thermoplastic polymers

In several polymers the regular sequencing of the repeat units allows elements of neighbouring chains to pack together in a preferred, lower energy, configuration. Such packing can be disrupted by mechancial work, but usually this is achieved by heating the polymer above its melting point. In the solid phase these locally ordered regions, or crystallites, act as physical crosslinks, preventing the dissolution of the molecular network in the presence of solvents. Crystallinity also enhances the high temperature performance of the polymer and, in particular, provides added resistance to long term phenomena such as creep under load. Useful semi-crystalline polymers usually contain between 5 and 50% of the polymer in the crystalline phase. If the crystallinity is too low, then the benefits of the physical crosslink network are not observed; if it is too high, the crystalline phase severely restricts the energy absorbing capability of the amorphous regions and the polymer may, in consequence, be brittle. The optimum level of crystallinity for a thermoplastic polymer to be used as a matrix for composites appears to be between 20 and 35%. Semi-crystalline thermoplastic polymers as matrices for high performance composites are a small, but rapidly growing, family.

The advantages of crystallinity are offset by some problems. The most obvious problem is how to preimpregnate the fibres with resins whose very nature prevents them from being readily dissolved and so amenable to conventional solution processing. Other difficulties arise from the close packing of the chains in the crystalline regions. This close packing usually means a large density change as the melt solidifies. The high density of the crystalline regions in comparison to the amorphous phase in which they are suspended means that those different regions will scatter light differently, so that the resin appears opaque: one polymer where this does not occur is poly (4-methyl pentene-1), where the crystalline and amorphous phases scatter light equally. In a highly filled composite system opacity of the matrix phase is not usually a cause for concern, although it can constrain some applications with transparent fibres. The differential shrinkage between amorphous and crystalline phases does, however, tend to confer a slightly mat surface finish to mouldings, and composite materials made with semi-crystalline matrices tend to have less satisfactory cosmetics than their amorphous cousins. One factor which does sometimes cause concern is that the level of crystallinity in such polymers can be varied by differences in processing history: rapid cooling from the melt causes low crystallinity; very slow cooling, or annealing near the melting point, may lead to excessive crystallinity. Preferred systems have a broad processing window within which the optimum crystallinity is achieved. The study of crystallinity in polymers is the subject of many scientific treatises[26], and, as a result, there is a large body of experience upon which to draw in resolving such issues.

Semi-crystalline polymers have order at several levels. Most evident are the families of crystallities that originate from a nucleation point and grow in a spherulitic fashion. The spherulitic structure is readily demonstrated and size can be changed by variations in processing history. Spherulites are not the primary determinants of the properties of semi-crystalline polymers: it is the level of crystallinity which is the most important factor.

2.1.5 Polymer blends and compounds

With such a broad spectrum of thermoplastic polymers with complementary properties from which to choose, it is natural to consider if optimization to function can be achieved by the use of polymer blends. The criteria on which the blend or compound are selected are varied and can include temperature performance, stiffness, ease of processing, and, not least important, cost. In general a blending process will increase one property at the expense of another; the most favourable indication for a blend is when it simultaneously enhances two properties. One example of such a synergy is the addition of a small quantity of liquid crystal polymer to a standard thermoplastic polymer[27]. This reduces melt viscosity of the host polymer and can also lead to increased stiffness. The level of environmental resistance offered by the best semi-crystalline polymers is sometimes more than is required for certain applications. Dilution of such resins with miscible amorphous polymers can lead to an increase in glass transition temperature, a broadening of the processing window and a reduction in cost. The blend route can also offer a means to impregnation. For example, polyphenylene oxide can be dissolved in styrene and impregnated into fibres, the styrene being subsequently polymerized to form a compatible blend[28]. Incompatible blends can also be of interest, especially if their microstructure can be controlled to an interpenetrating network where the advantageous properties of both components can be exploited. This approach has been particularly successful in blends of thermoplastic and thermosetting systems[29]. Such blends may also permit preferred wetting of the fibres by one phase that physically links the non-wetting resins to the structure. The addition of fine particle filler can increase stiffness and give enhanced crystalline nucleation. Such is the cost of developing a wholly new polymer to meet a new application, and such are the diversity of properties that can be obtained by compounding, that we may expect to see considerable effort deployed in this area in future.

2.1.6 The 'Victrex' range of aromatic polymers

The 'Victrex' range of polymers from ICI provides a series of high performance engineering materials whose origins, history and applications have been described by Rose[30] and Belbin and Staniland[31]. Polyethersulphone (PES) and polyetheretherketone (PEEK) are the best known representatives of this family, whose members are based on separating rigid aromatic units:

$$Ar_1 \quad Ar_2 \ldots \ldots \ldots Ar_m$$

with either flexible

$$f_1 \quad f_2 \quad \ldots \ldots \ldots f_q$$

or stiff

$$s_1 \quad s_2 \quad \ldots \ldots \ldots s_p$$

linkages. There are also two end groups, thus:

$$p + q = m-1$$

In homoploymers the rigid aromatic units and the flexible and stiff linkages are arranged in a regular sequence to provide the repeat unit. A repeat unit involving two similar aromatic units (Ar) and one flexible (f) and one stiff (s) linkage would be written:

$$- [Ar - f - Ar - s] -$$

The homopolymer from such a repeat unit would be written:

$$[Ar - f - Ar - s]_n$$

where n is the number of times that this sequence is repeated.

It is also possible to produce a range of copolymer materials: these usually have a random sequencing of two or more different repeat units.

Although it is possible to produce cyclic oligomers of this family[32], the chains are, in general, not endless but are terminated at some point by an end group. The usual end groups found in the 'Victrex' family of polymers are fluorine in PEEK, and chlorine in PES.

The properties of the polymer depend upon the following factors: the rigid aromatic units, the flexible and stiff spacers, the number of times the monomer sequence is repeated, and the end groups. The end groups are usually intended to be inert, but active ends can be used to achieve special chemical effects, in particular to react either to form a crosslinked polymer or to interact with another resin system[33]. The rigid aromatic ring structures are the backbone building blocks of the 'Victrex' family. These ring structures have outstanding chemical stability and also provide a high carbon content, so that, in fires, the resin will char. Most commonly the aromatic unit is a phenylene ring:

but various double ring structures are also used if extra stiffness and high temperature performance is required. Further, special function can be given to the polymer by modification of the rigid rings. For example, by sulphonation, the polymer can be induced to become permeable to water, allowing it to be considered as a membrane material. In general such modification is not desirable in composite materials, and the preferred structures are simple phenyl or biphenyl rings.

The flexible linkages are the key to inducing some freedom for the chain to rotate, thereby allowing it to melt and be processed. The most widely used flexible linkage in this family of resins is the ether link:

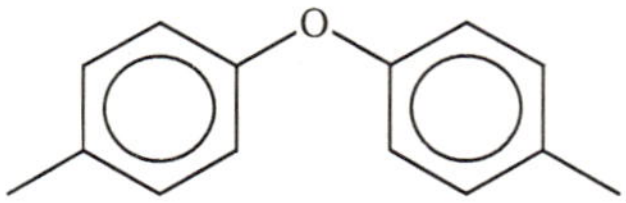

There are also stiffer linkages that provide some mobility. In the case of polyethersulphone this linkage is a bulky sulphone group and there is one such group for each ether linkage:

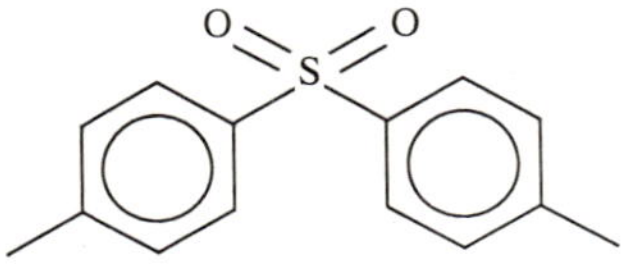

For polyetheretherketone there is one stiff ketone linkage for every two flexible ether linkages.

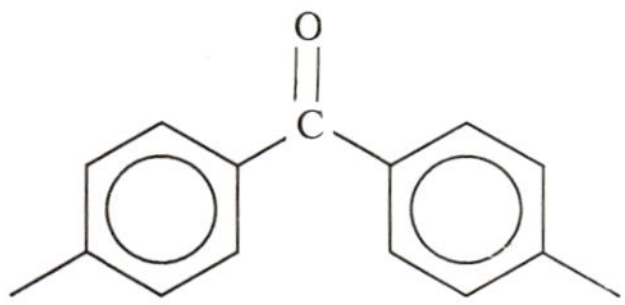

In comparison to the sulphone group the ketone link is fairly compact. Another feature of the ketone link is that the angle which it makes with the two-ring structures, 125°, is essentially the same as that of the ether linkage. The sulphone group provides a narrower angle. Because of the bulkiness of the sulphone group and improbability of matching up ether groups in a chain that is inevitably locally curved, it is extremely difficult to crystallize the polyethersulphone family. By comparison, the compatibility of the ether and ketone linkages means that it is possible for the chain to be locally straight and for adjacent molecules to come into register without identical sequencing:

This allows for easy crystallization of the polyetherketone family[34].

As well as the simple primary members of the 'Victrex' family, polyether-sulphone and polyetheretherketone, there are a range of special materials, some of which have been designed especially as matrices for composite materials.

2.1.7 Polyetheretherketone

Staniland[35] made an extensive review of the polymerization chemistry for the class of resins known as poly(aryl ether ketone). He concentrated on the first freely available polymer whose systematic name is

poly(oxy-1,4-phenyleneoxy-1,4-phenylenecarbonyl-1,4-phenylene)

amply justifying its abbreviated form polyetheretherketone and acronym PEEK. This polymer is produced commercially by a nucleophilic process in a dipolar aprotic solvent. More specifically the ingredients hydroquinone, 4,4' difluoro-benzophenone and potassium carbonate are reacted together in diphenylsulphone in the temperature range 150–300°C to form polyetheretherketone, potassium fluoride, carbon dioxide and water:

The polymer is subsequently isolated, but small residues of diphenylsulphone, which at normal temperatures is a crystalline solid, are sometimes detectable in the resin.

The single most important parameter describing a polymer is its molecular weight. This is usually characterized in dilute solutions, but, for PEEK, the only known common solvent is concentrated sulphuric acid, which may also interact chemically with the chain. Satisfactory solutions have been achieved in mixtures of phenol and trichorobenzene at 115°C. Devaux and his colleagues[36] have developed techniques based on gel permeation chromatography to characterize these polymers. This method has been used as a standard test for seven years and is regarded as a satisfactory system for polyetheretherketone. The description of molecular weight and its significance in composite matrices is considered in Appendix 4.

High molecular weight leads to high resin toughness but also high viscosity. The optimization of the resin calls for a balance between these properties. Because of the need to have some flow of the resin to aid the wetting out of the fibres and the processing behaviour, and because extremes of matrix toughness are not obviously rewarded by further increases in composite toughness[37], that compromise has generally been resolved towards the lower end of the range of molecular weights, which are generally considered to give useful service performance. Particularly outstanding combinations of toughness and processability can be achieved in matrix resins of narrow molecular weight distribution. Figure 2.1 shows a typical molecular characterization of PEEK extracted from a sample of composite material. In this case the weight average molecular weight is 30,000 and the number average molecular weight is 13,000, giving a ratio close to the theoretical optimum for this type of polymerization.

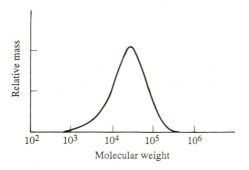

Figure 2.1 Molecular weight characterization of PEEK extracted from a composite material

The implication of molecular weight can be further appreciated by considering the geometry of the chain. For polyetheretherketone the atomic combination of the repeat unit is nineteen carbon, twelve hydrogen and three oxygen atoms, allowing for a total molecular weight of 288.

From this we can judge the degree of polymerization (DP) in each chain:

$\overline{M}_W$ = 30,000 DP_W = 104 (weight average degree of polymerization)

$\overline{M}_N$ = 13,000 DP_N = 45 (number average degree of polymerization)

From the basic chemical structure a model of the PEEK molecule can be developed[38]. From this we can deduce the length of the monomer unit as 1.5 nm, so that the fully extended length of a chain whose molecular weight is 30,000 (DP_W = 104) is 156 nm.

From the same model it is possible (Appendix 5) to deduce the effective cross sectional area of the chain, and so an effective diameter. In the case of PEEK this diameter is 0.57 nm. From this we can deduce that the aspect ratio of a chain whose molecular weight is 30,000 is 273.

The chain is not usually totally straight: it has the ability to rotate at the ether and ketone linkages. This flexibility allows the chain, at rest, to take up a random coil configuration. The volume which that chain occupies can be approximately defined by the radius of gyration of the molecule (Appendix 6). In the case of PEEK, where the bond angles are $125°$[39], we can deduce a radius of gyration (r) of 16.7 nm for a molecule of molecular weight 30,000. The volume occupied by such a molecule is then:

$$\text{Space occupied,} \frac{4}{3}\pi r^3 = 19,500 \text{ (nm)}^3$$

Within the same space are numerous other molecules. We have deduced the diameter (D) and length (L) of such a chain as 0.57 nm and 156 nm respectively. From this the volume of the chain can be deduced:

$$\text{Volume of chain,} \frac{\pi D^2}{4} L = 40 \text{ (nm)}^3$$

The ratio of the space occupied by the chain to the volume of the chain gives the number of chains of equal length which can be fitted into the same volume. Of course the chains do not necessarily integrate in that way. Most chains occupy spaces which overlap (Figure 2.2, overleaf). Further, the interaction will not only be with chains of the same length but will include all lengths present. The ratio gives a feel for the level of interaction between chains and, for a molecular weight of 30,000, this level of interaction is approximately 500.

Each molecule interacts with a large number of other molecules. Many of the

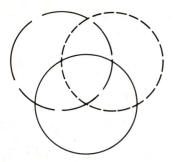

Figure 2.2 Overlapping spaces occupied by separate chains

properties of a polymer depend upon the level to which the chains entangle. The entanglement of chains depends on their tortuosity. One index of tortuosity is the molecular length divided by twice the radius of gyration. For a PEEK chain of molecular weight 30,000 this entanglement parameter is 2.3.

Because we write on flat sheets of paper the PEEK molecule:

usually appears flat. This convenient projection is not completely true even in crystal packing where, although the oxygen atoms all lie in a single plane, phenylene rings are inclined so that their director planes are at an angle of approximately ±37° alternately down the chain[40].

The full detail of the crystalline morphology, originally discussed by Dawson and Blundell[41], has been extensively reviewed[42,43] and continues to be the object of refinement. Appendix 5 includes the main conclusion of King and his colleagues[44]. The main crystal texture is formed by the crystalline lamellae (Figure 2.3). The

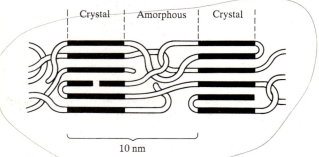

Figure 2.3 Crystalline morphology (schematic)

periodicity of the lamellae is about 10 nm, with the thickness of the lamellae depending on the level of crystallization but usually in the range 25–40% of this spacing[45]. The periodicity of the crystal lamallae must be considered in connection with the molecular dimensions. Clearly a fully extended chain of molecular weight 30,000 with a length of 156 nm could extend through a dozen lamellae. However, if the chain is crystallized from an unperturbed state, the usual case for composite matrices, then such a chain would only have a dimension of about 30 nm (twice the radius of gyration – Appendix 6) and so would only be expected to participate in two or three lamellae. Participation in two such lamellae (Figure 2.4), or forming a loop through which other chains of adjacent lamellae are threaded (Figure 2.5), provides effective physical links in the network. Such links are required in order to develop the full properties of the material.

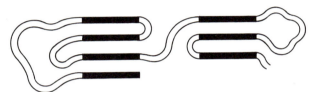

Figure 2.4 Molecule linking between crystals (schematic)

Figure 2.5 Threaded molecules linking adjacent lamellae (schematic)

The detail of how the material is ordered in both the crystalline and amorphous regions continues to be an area of active controversy. For the most part there is general agreement that the chains in the crystalline lamellae can pack with the ether and ketone groups in any register. There is, however, some evidence[46] that slightly different crystallization patterns are observed during very slow crystallization processes when perhaps there is more opportunity to achieve perfect register of the ketone groups. There is also discussion of order and even a pseudo-crystalline state[47] in the so-called amorphous regions. Some anomalies remain to be resolved. The equivalence of the packing of the ether and ketone groups should allow all members of the poly(aryl ether ketone) family to be isomorphous: PEEK is miscible with PEK but not with PEKK, while PEKK is miscible with PEK and PEEKK[48]. These features suggest an ample field for future research to obtain a full understanding.

The level of crystallinity achieved in PEEK polymer depends on the processing history[49-72]. Very rapid cooling can produce an amorphous polymer. This can subsequently be annealed to achieve any desired level of crystallinity. Molten PEEK suddenly cooled to 220°C will crystallize in about 6 seconds. There is a broad temperature range, 190°C to 260°C, where crystallization will occur in less than 10 seconds. Outside this range crystallization is progressively slower, taking about 1 minute at 180°C or 290°C and 10 minutes at 160°C or 320°C. Crystallization at above 300°C should be avoided if possible, since this can lead to very high levels of crystallinity, which may unduly constrain the amorphous regions between lamellae, so compromising the toughness of the material. Accordingly it is desirable to cool the material from the melt sufficiently rapidly to avoid crystallization in that temperature range. The optimum level of crystallinity for PEEK resin is 25 to 40%.

Crystallization behaviour can also be affected by the presence of nucleating agents, including graphite. Local stresses may orient the chains, providing just sufficient local order to initiate the process. One factor that produces a perturbation of the molecule, resulting in local stress, is the crystallization process itself. This causes crystals to grow in families radiating from one point. Such families are known as spherulites. The most potent nucleating agents are residual unmelted seeds from previous crystallites. To eliminate these completely it is necessary to raise the melt temperature to at least 360°C, 25°C above the nominal melting point.

Crystallization behaviour is conveniently studied by differential scanning calorimetry[73]. In this process the specific heat, Cp, is measured as a function of temperature both on heating and cooling. Care must be taken in the interpretation of such data, but, with appropriate controls, it can provide useful information about the polymer and also evidence of the previous history of that material.

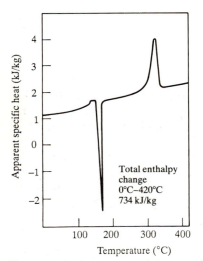

Figure 2.6 DSC heating trace for amorphous PEEK

Figure 2.6 shows the apparent specific heat of amorphous PEEK resin during a heating cycle from 0°C to 420°C at a heating rate of 20C/min. There is a small, clearly defined increase in specific heat at the glass transition temperature. This is followed by a dramatic exotherm as the sample crystallizes at about 170°C. That crystallinity subsequently melts over the temperature range 300° to 350°C, with the peak melting rate at about 340°C. The heat contents of the exotherm and subsequent endotherm are equal.

The same thermal history for a sample of semi-crystalline PEEK polymer is shown in Figure 2.7. The change in specific heat at the glass transition temperature is shifted to a higher temperature and is less clearly marked. At about 270°C an initial melting and recrystallization process is evident. This indicates that the sample was originally crystallized about 260°C. The subsequent high temperature melting process is similar to that of the amorphous sample.

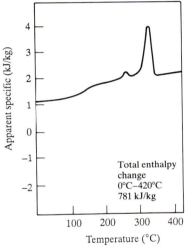

Figure 2.7 DSC heating trace for semi-crystalline PEEK

The total enthalpy change up to 420°C for the semi-crystalline sample is 781 kJ/kg and for the amorphous sample is 734 kJ/kg. The difference in these figures divided by the latent heat of crystallization of PEEK polymer (130 kJ/kg) indicates that the semi-crystalline sample contained about 35% crystallinity.

Beyond the scale of crystallinity are the families of crystalline entities that initiate from a nucleation point and grow in an approximately spherical manner before impinging with their neighbours (Figure 2.8). The growth pattern of such spherulites has been reviewed by Medellin-Rodriguez and Phillips[74], who suggest that their maximum radial growth rate occurs at about 230°C with a speed of about 0.2 μm/s.

The protospherulite is actually a sheaf like structure[75-77] rather than a point. Spherulitic texture can be revealed by etching techniques[78], which preferentially

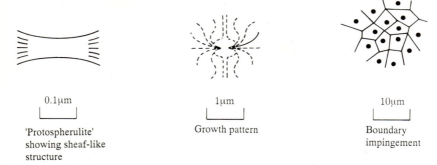

<table>
</table>

0.1μm	1μm	10μm
'Protospherulite' showing sheaf-like structure	Growth pattern	Boundary impingement

Figure 2.8 Spherulite textures (schematic)

dissolve the amorphous material. They are also seen in polarized light and are dramatic in appearance. This dramatic appearance belies their importance: the level of crystallinity is more important than the size of the spherulite. It is only during extremely slow cooling processes in a constrained situation that there will be interspherulitic voids and weakness. In PEEK the normal spherulite size varies from 1 to 10 μm or even larger, dependent on process history and nucleation density. In the absence of nucleation slowly crystallized materials will tend to have large spherulites.

A summary of the morphology observed in polyetheretherketone is provided in Table 2.1.

Table 2.1 Morphology of PEEK

Molecular weight		30,000
Chain diameter (nm)		0.57
Chain length (nm)		156
Radius of gyration (nm)		16.7
Volume of chain (nm)^3		40.0
Space occupied by chain (nm)^3		19,500
Interaction parameter*		500
Tortuosity parameter†		2.3
crystal unit cell (nm)	a axis	0.783
	b axis	0.594
	c axis	0.986 along the molecular axis
Crystal lamella thickness (nm)		1 to 4
Crystal lamella spacing (nm)		about 10
Spherulite size (nm)		1,000 to 10,000 or above
(μm)		1 to 10 or above

* Space occupied by chain divided by volume of chain.
† Chain length divided by twice the radius of gyration.

Despite its high melting point, PEEK is an outstandingly stable polymer. Provided that oxygen is excluded, the melt is thermally stable for 1 hour or more at 400°C. In the presence of oxygen, or at higher temperatures and longer times, some chain branching or crosslinking can occur, causing an increase in melt viscosity and a reduction in the ability to crystallize.

The thermophysical properties of PEEK (Table 2.2) are essentially as would be expected of a semi-crystalline polymer.

Table 2.2 Thermophysical properties of peek

	@23°C		@380°C
Glass transition temperature (Tg)	143°C		
Maximum continuous service temperature	250°C		
Melting point	334°C		
Density: amorphous (kg/m³)	1,264		1,080
20% crystalline (kg/m³)	1,291		
40% crystalline (kg/m³)	1,318		
fully crystalline (theory) (kg/m³)	1,400		
Latent heat of fusion (100% crystalline) (kJ/kg)	130		

| | *Temperature range* | | |
	23–143°C	*143–334°C*	*334–420°C*
Specific heat (kJ/kg°C)	1.1–1.5	1.6–2.1	2.1–2.2
Thermal conductivity (W/m°C)	0.25		0.35
Thermal diffusivity (m²/s)	0.18×10^{-6}		0.15×10^{-6}
Coefficient of thermal expansion (/°C)	47×10^{-6}	108×10^{-6}	120×10^{-6}
Equilibrium water content	0.5%		

For a structural material it is the mechanical properties which are of pre-eminent concern. There are two major transitions in the mechanical response of PEEK as a function of temperature: at 143°C, usually referred to as the glass transition temperature, and at the melting point, 334°C. Dynamic mechanical analysis is a convenient way of displaying these transitions (Figure 2.9).

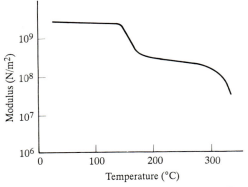

Figure 2.9 Dynamic mechanical analysis of PEEK

The modulus of PEEK in the usual service temperature range, $-60°$ to $+120°C$, is about $3.5\,GN/m^2$, falling to one-tenth of this value above the glass transition temperature (Tg). Although, for structural applications, a service temperature of about 120°C would appear to be the maximum, crystalline resins are self-supporting up to the melting point, and there are many applications where this resin gives excellent service well above Tg. The mechanical properties of PEEK resin at ambient temperature are outlined in Table 2.3.

Table 2.3 Typical mechanical characterization of PEEK at 23°C

	Modulus	Strength
Uniaxial tension	$3.6\,GN/m^2$	$92\,MN/m^2$
Uniaxial compression	$3.6\,GN/m^2$	$119\,MN/m^2$
Simple shear	$1.3\,GN/m^2$	$\sim 60\,MN/m^2$
Bulk	$6.2\,GN/m^2$	
Poisson's ratio	0.40	
Fracture roughness		
K_{IC}		$4.8\,MN/m^{3/2}$
G_{IC}		$6.6\,kJ/m^2$
Rockwell hardness		
M scale		126
R scale		99
Coefficient of friction		0.27

In addition to this basic characterization, the long-term creep, fatigue and tribological properties have been studied in detail[79,80] and can be described as excellent in the context of other engineering polymers.

Table 2.4 Reagents having no significant effect on PEEK after seven days' exposure at 23°C (unless otherwise stated)

Acetic acid	Toluene
Acetone	methyl ethyl ketone
Benzene	Ethylene glycol
Carbon tetrachloride	Xylene
Diethyl ether	Benzaldehyde
Dinethyl foramide	Gasoline
Ethyl acetate	Concentrated ammonium nitride
Ethyl alcohol	28% hydrogen peroxide solution
Heptane	30% sulphuric acid
Kerosine	40% nitric acid
Methyl alcohol	Water (at 95°C)

In respect of resistance to hostile environments, PEEK is generally considered to be outstanding in the field of polymeric resins. The only common material that will dissolve PEEK is concentrated sulphuric acid. Concentrated nitric acid does not dissolve PEEK but does cause it to yellow and significantly degrades tensile strength. PEEK will absorb between 2–5% of formalin, concentrated sodium hydroxide, 40% chromic acid and concentrated hydrochloric acid; but these cause no significant change in tensile strength. Reagents which appear to have no significant effect on PEEK after seven days' exposure are listed in Table 2.4. Besides outstanding resistance to chemical reagents, PEEK has exceptional resistance to radiation[81] and a V-O flammability rating down to 1.45 mm thickness, with exceptionally low smoke and toxic gas emission[82].

PEEK is commercially available in a variety of forms. There is neat resin in powder or granule form. 'Victrex' PEEK is available in fibre (Zyex) and film (Stabar) forms, and a range of filled extrusion and moulding compounds, besides continuous fibre reinforced products. This broad range encourages the use of the material forms in combination and the reclaim of product forms from one application to another: thus PEEK powder, fibre or film are variously used as ingredients of composite materials. PEEK film may be used in bonding processes or to give optimized surfaces to the moulding; continuous fibre reinforced PEEK can be used for selective reinforcement of mouldings formed from neat resin or filled compounds; and offcuts from structural composite mouldings can, with the addition of extra resin, be ground down to provide high value moulding compounds. We are only beginning to exploit the versatility and compatibility of thermoplastic product forms.

2.2 Reinforcing fibres

Reinforcing fibres are the backbone of structural composite materials. The high stiffness and strength of these fibres provide the characteristic mechanical properties of advanced composites. The main criteria for fibres to be used in thermoplastic structural composites are that they should be available in long or continuous fibre form; be stiff; with moduli in excess of $50 \, GN/m^2$; have good strength; exhibit resistance to solvents; and be resistant to the temperatures of processing, which can be up to 400°C. Since structural composite materials are often used in weight sensitive applications, low density is desirable, so that the composite has high 'specific' properties, i.e. high stiffness per unit weight. The most specific property of concern to the user is, most often, performance per unit cost and, since the fibres represent the major part of such composites, their cost is also a major factor. A process of elimination leaves three main categories of interest: organic fibres based on rigid, aromatic polymers; inorganic fibres; and carbon or graphite fibres. High stiffness and strength imply strong interatomic and intermolecular bonds and few strength limiting flaws. These factors are achieved in carbon and organic polymer fibres by their highly orientated structures: in the amorphous, inorganic fibres, such as glass, they depend on a rigid, three-dimensional network morphology. A comparison of key properties for the three classes of fibre is listed in Table 2.5.

Table 2.5 Typical fibres used in structural composites

Fibres	Density kg/m^3	Modulus GN/m^2	Strength MN/m^2	Diameter μm
ORGANIC FIBRES				
Aramid	1,450	124	3,600	12
INORGANIC FIBRES				
E Glass	2,600	73	3,400	17
S-2 Glass	2,490	87	4,500	10
γ-Alumina	3,250	210	1,800	20
CARBON FIBRES				
High strength	1,780	227	3,600	7
Intermediate modulus	1,800	303	5,500	5
Ultra high modulus	2,150	724	2,200	10

The manufacture, structure and properties of these fibres are reviewed in books by Bunsell[83] and by Watt and Perov[84]. The field of high performance fibres for structural composites has evolved rapidly during the last two decades, and that pace shows no sign of slackening.

2.2.1 Organic polymeric fibres

The most widely used organic fibres in structural composites are the aramid fibres spun from liquid crystalline solutions of poly(paraphenylene terephthalamide). The conventional fibres of this family have a modulus of about 124 GPa, which, combined with their low density (1,450 kg/m^3), gives good specific stiffness and high specific tensile strength. Aramid fibres with moduli up to 180 GPa have been reported[85]. Research into other liquid crystal polymer fibres, in particular poly(paraphenylene benzobisthiazole), has indicated that moduli up to 330 GPa can be achieved[86]. In spite of their high strengths and stiffnesses in tension, highly oriented polymeric fibres are relatively weak in compression and torsion because of their microfibrillar structure[87]: this constrains the range of structures where such fibres can be used. A second weakness, in the context of thermoplastic matrix systems, is the potential to degrade the strength of the fibres by the high temperatures of the processing operation. This problem is most evident with ultra oriented polyethylene fibres, where it would be necessary to limit the choice of matrices to those having melting points below 140°C, which would severely constrain the serviceability of the composite. Even the aramids show some evidence of degradation after processing at temperatures of 350°C, so that their use has mainly been limited to conjunction with amorphous polymer matrices, which can be prepregged from solution and fused at about 300°C. Khan[88] has prepared satisfactory composites based on aramid fibres in a low melting point polyketone matrix. Aramid fibres are important reinforcing fibres for thermoplastic matrices, finding their best expression in applications where impact resistance is critical but compression performance is not. The heavy investment in the science of liquid crystal polymers during the last decade promises further improvements of this class of fibre in future.

2.2.2 Inorganic filaments

The most widely used inorganic fibres are glass fibres prepared by spinning from a melt of mixed oxides. The fibres have a rigid polyhedral silica based structure, and are commonly non-crystalline and isotropic. Conventional E-glass fibres are made in large tonnages for a variety of applications and, as a result, are inexpensive relative to other reinforcing fibres. This makes them a natural first fibre of choice in any application. Somewhat superior in properties to conventional E-glass fibres are systems known as R- and S-glass. Made in smaller quantity, these fibres are inherently more expensive than their mass produced cousins, but the cost of prepregging each fibre is the same. For thermoplastic composites the improved cosmetics of these higher performance fibres may actually lead to easier prepregging; thus the price differential in composite form is not so great. Glass fibres are very flaw-sensitive and their strength degrades considerably after drawing as surface flaws develop. A protective coating or 'size' is therefore applied immediately after drawing to minimize damage. This size is generally polymeric in nature and can also be designed to enhance adhesion between the glass surface and polymer matrix. The high processing temperatures associated with thermoplastic polymers may call for special size systems that are not available in the mass produced E-glass fibres but can be readily tailored on to their higher-performance cousins. For high performance structural composites it is these speciality glass fibres which tend to be preferred: the development of PEEK/glass composites is traced by Turner[89] and Hoogsteden[90]. The isotropic structure of glass fibres means that they do not show any pronounced weakness in compression or torsion, in comparison with fibres made by very high molecular orientation processes. Their high strain to failure also makes them competitive with aramid fibres where energy absorption is a consideration. An early use of glass fibres in structural composites was because of their dielectric properties, which made them good materials for radomes[91]. The increasing sophistication of applications, where structural performance alone is not enough, makes increasing demand for speciality glasses. Glass is the oldest of the family of reinforcements for structural composites; it is also a system whose versatility will ensure its continued importance.

Glass fibres have two significant problems in structural materials: they are heavy, and it is difficult to obtain good adhesion between the matrix and the fibre. The problem of high density is a feature of all the inorganic fibres, and there is little that one can do to overcome it. The problem of adhesion at the interface of glass fibres can be attacked by the use of appropriate coupling agents incorporated into the protective size (see Section 2.3). Further, the hydrophilic nature of the silanol groups on the surface of glass makes them water sensitive. The development of an optimum size system for thermoplastic that will counteract this tendency and not degrade under high temperature processing represents a considerable challenge.

Boron monofilaments were one of the earliest reinforcing agents for structural composites[92], but high cost relative to carbon fibres has severely restricted their use. The outstanding property of boron fibres is their high compressive strength, in part a function of their large diameter (typically 140 μm). This large diameter makes them particularly difficult materials for prepregging and processing, since only a very gentle radius of curvature can be tolerated. However, that same large diameter gives a relatively low fibre surface area per unit volume. This means that they can be readily wetted with thermoplastic resins. Another feature of the large diameter of boron fibres is that the spaces between the fibres are large (20–100 μm), allowing the matrix phase itself to be a composite material reinforced with conventional small diameter fibres. The potential for thermoplastic hybrid composites of this kind has yet to be fully explored.

Recently there has been a resurgence of interest in ceramic fibres, with, for example, continuous alumina and silicon carbide fibres becoming available commercially. With temperature stabilities of over 1,000°C, they are of particular use for reinforcing metals. Continuous alumina fibres have also been used for reinforcing thermoplastic composites[93], and both silicon carbide and silicon nitride fibres are making their debut as components in polymeric composites[94]. Although currently more expensive than glass or carbon fibres, these ceramic fibres extend the range of properties available for inorganic fibres; they can combine good dielectric properties with high stiffness. A summary of recent developments in inorganic fibres is given by Bracke, Schurmans and Verhoest[95].

2.2.3 Carbon fibres

High performance carbon fibres were first produced in the early 1960s by Shindo in Japan and Watt at the Royal Aircraft Establishment in England[96,97]. These fibres were prepared by the carbonization of polyacrylonitrile (PAN), as are most of the fibres available commercially today. Carbon fibres have also been made from a variety of other precursors, such as rayon and pitch. The preparation, structure and properties of carbon fibres have been the subjects of many reviews[98–102].

Carbon fibres have been shown to consist of intermingled fibrils of turbostratic graphite, with basal planes tending to align along the fibre axis in a crimped or contorted fashion[103]. This highly anistropic morphology gives rise to moduli in the range 200 to 900 GN/m² parallel to the fibre long axis and around 15 GPa in the normal direction, comparing with 1,060 GN/m² and 37 GN/m² for a single crystal of graphite along and normal to the basal plane direction respectively[104]. Ultra-high moduli are achieved in fibres prepared from liquid-crystalline mesophase pitch: the higher degree of orientation in the precursor translates through to the final carbonized fibre, leading to larger and more orientated graphite crystallites. Pitch fibres have more regular internal structures than PAN fibres, with the basal planes tending to orient in a sheaf-like, spoke-like or onion skin array[105,106].

Carbon fibres behave as brittle materials and their strengths depend on their internal structures and the presence and distribution of flaws and defects. In general the higher the modulus, the lower the strength. The main drive of the manufacturers over the past few years has been to produce fibres in the intermediate modulus range with improved strengths. This is usually achieved by improving the quality of the precursor fibres and further stretching of fibres during production.

In addition to mechanical properties, there are a number of other characteristics of carbon fibres of interest to the composite engineer. Carbon fibres have a low, actually slightly negative, coefficient of thermal expansion, typically around -1×10^{-6} per °C along the fibre axis[107]. The resulting low thermal expansions in carbon fibre reinforced composites are of interest in structures that require high dimensional stability. Other properties include high thermal and electrical conductivity[108]. The potential to tailor modulus and strength has made carbon fibres the most widely used reinforcing system in high performance structural composites.

2.2.4 High strength carbon fibres

The most widely used carbon fibres are the family of high strength fibres, with modulus of 230 GN/m² and strength 3,600 MN/m². Courtaulds XAS-0 and Hercules AS4 are the fibres used with PEEK resin in the commercialized grades of APC-1 and APC-2 respectively.

Typical dimensions of a high strength carbon fibre are listed in Table 2.6 and can be compared with those of the resin previously shown in Table 2.1.

Table 2.6 Typical dimensions of high-strength carbon fibre

	Dimensions
Graphite plate spacing	0.35 nm
Locally ordered family of plates thickness	~5 nm
Graphite plate width	~3 nm
Fibre crenellation width	0.5 μm
Fibre crenellation depth	0.2 μm
Fibre diameter	~7.1 μm
Depth of fibre tow in prepreg	~0.1 mm
Width of tow in prepreg	~10 mm
Tow length in prepreg	~100 m
Tow length as made	~5 km

In addition to the surface crenellations, carbon fibres have some internal porosity at the level of about 10%. These pores are disclinations in the graphite plate structure and may be on a scale of about 10 nm: larger pores would lead to reductions in fibre strength. Such features do not appear to be readily accessible from the fibre surface, which is usually sealed in the surface treatment processes.

Not least of the advantages of carbon fibre as a reinforcing fibre for composite materials is the ability to tailor the surface activity to make a strong interface with the matrix. Methods of activating the surface of carbon fibres to achieve this are discussed in Section 2.3.

In the final analysis it is the stiffness that determines the utility of carbon fibre in composite materials. However, the structure of a carbon fibre is highly anisotropic, and the stiffness depends on the mode of deformation. While the tensile stiffness along the fibre direction can be measured with considerable precision, the transverse and shear properties are less easy to determine. Following the work of Rogers and colleagues[109] and Wagoner and Bacon[110], we may estimate the stiffness properties of AS4 type high strength fibres as follows:

Stiffness and strength

E_1	Axial tensile modulus (GN/m²)	227
E_2	Transverse tensile modulus (GN/m²)	15
G_{12}	Axial shear modulus (GN/m²)	20
G_{23}	Torsional shear modulus (GN/m²)	5
σ_1	Axial tensile strength (MN/m²)	3,650

Poissons ratio

ν_{12}	Transverse contraction with axial extension	0.25
ν_{23}	Transverse contraction with transverse extension	0.40
ν_{21}	Axial contraction with transverse extension	0.013

where:

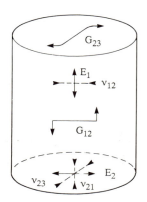

Thermal properties[99, 110, 115])

α_1	Coefficient of axial thermal expansion (per °C)	-1.2×10^{-6}
α_2	Coefficient of transverse thermal expansion (per °C)	12×10^{-6}
K_1	Thermal conductivity along the fibre (W/m°C)	~ 16
K_2	Thermal conductivity across the fibre (W/m°C)	~ 3

Cp Specific heat (J/kg°C)

@23°C 0.75
@143°C 0.99
@380°C 1.42

ρ Density (kg/m³)

1,780

These estimates are particularly uncertain in respect of the transverse and shear properties. The transverse thermal expansion quoted is by Sheaffer[111], who used a direct measurement of individual fibres by means of laser diffraction. Indirect measurement, whereby this value is back calculated from measurements on composite materials, sometimes suggest values approximately double this figure[112], but those calculations may assume incompressibility of the fibre and matrix, and ignore the volume changes due to internal stress. Some estimate is necessary in order to recognize that, while the axial stiffness of the fibre is nearly two orders of magnitude greater than that of the resin, the transverse properties are very much closer. Precise measurement of these properties is highly desirable in order to achieve a full micromechanical understanding of the composite behaviour.

2.3 Interfaces and interphases

Much of the early history of composite materials sought to optimize the toughness by deliberately designing a weak interface between matrix and reinforcement[113]. The theory was that a crack propagating through a brittle matrix would be deflected at the surface of the reinforcing fibre and debond that fibre from the matrix creating a large amount of free surface and thereby absorbing energy. Without a weak interface the crack would propagate directly through the fibre leading to catastrophic failure. The weak interface theory is still applied today with brittle matrix composites but, with thermoplastic polymers, we have a means of dissipating energy within the matrix. By utilizing this mechanism it is possible to design a tough composite material with very strong adhesion at the interface[114] (Figure 2.10). Lustiger[115] emphasizes that, in the commercially available carbon fibre PEEK product APC-2, there are no conditions of loading that destroy that

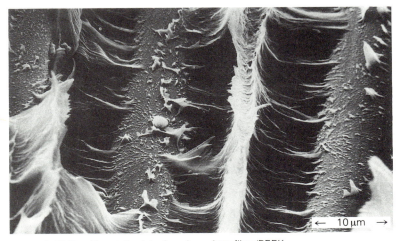

Figure 2.10 Ductility at the interface in carbon fibre/PEEK

interface and leave bare fibres, but he also points out that this is not a feature of other thermoplastic systems.

Typical properties of two composites made from PEEK resin, with two fibres of the same stiffness and strength but treated so that one fibre gave optimized adhesion and the other gave virtually zero adhesion are shown in Table 2.7.

Table 2.7 The influence of fibre matrix adhesion on the mechanical properties of high strength carbon fibre/PEEK composites (60% by volume carbon fibre), based on the results of Fife[116] and Barnes[117]

	Optimized adhesion	Virtually zero adhesion
Axial tensile modulus (GN/m^2)	145	140
Axial tensile strength (MN/m^2)	2,100	1,400
Coefficient of thermal expansion ($\times 10^{-6}$/°C)	0.24	0.24
Axial compressive strength (MN/m^2)	1,200	1,100
Transverse flexural strength (MN/m^2)	150	50
Interlaminar fracture toughness (kJ/m^2)	2.5	1.0

Note that the fibre dominated properties of stiffness, strength and thermal expansion are only slightly affected. This suggests that the resin makes a good shrink fit on to the fibre, but, as soon as there is an attempt to pull the resin away from the fibre, in transverse flexure or in delamination, the advantages of strong adhesion are obvious. Further, because the interface is not required to fail, it can be overdesigned, so that there is little prospect of some unexpected combination of circumstances causing a failure. The ability, with tough thermoplastic polymers, to design with a strong interface at once reduces the potential variability of the system, leading to enhanced quality assurance.

Having defined the desirability of achieving a strong interface, we must explore how this can be achieved. A variety of mechanisms have been proposed to account for adhesion within the fibre-matrix interphase. These include wetting, chemical bonding, mechanical and crystalline interlocking. Much of the art of manufacturing high quality thermoplastic composites lies in exploiting these mechanisms.

A necessary preliminary to achieving good adhesion is to impregnate the reinforcement with the matrix, thereby placing the resin and fibre in close physical juxtaposition. This central issue to the manufacture of composite materials is considered in Chapter 3. Before we reach that stage, we must first prepare the two surfaces to be joined.

2.3.1. Wetting of the fibre by the resin

An effective impregnation process allows the resin, either molten or in solution, to come into contact with the surface of every fibre. This is aided if the surface energetics of fibre and matrix are favourable, such that the contact angle between them is close to zero. This means that the surface energy of the fibre must be greater than the surface energy of the matrix. In addition, the smaller the interfacial surface energy, the greater compatibility between fibre surface and matrix[118]. A particular discussion of the fibre surface energetics and fibre matrix adhesion in carbon fibre/PEEK is given by Hodge, Middlemiss and Peacock[119].

The surfaces of carbon fibres are particularly amenable to tailoring. Having emerged from the final graphitization oven, their surfaces are chemically inert. Those surfaces can be activated in may ways[120]. The method most frequently used commercially is electrolytic oxidation. The surface energetics and surface atomic compositions of some untreated carbon fibres with a non-optimized surface, and of some carbon fibres of the same batch that have been given an oxidative surface treatment to promote adhesion by a thermoplastic matrix, are listed in Table 2.8[121].

Table 2.8 Surface energy analysis data for some carbon fibres, with and without an oxidative surface treatment[122]

Fibre surface treatment level	γ^d mJm^{-2}	γ^p mJm^{-2}	$\gamma^T = \gamma^p + \gamma^d$ mJm^{-2}	Surface elemental composition (atomic %) O	N
Untreated	27.2	8.5	35.7	2.2	2.1
Treated	26.1	22.3	48.4	7.3	4.0

where γ^d is due to the non-polar or dispersive interactions
$\quad\quad\gamma^p$ is due to the polar interactions
$\quad\quad\gamma^T$ is the total surface energy.

The fibre surface treatment doubles the polar component of the surface energy, and the level of surface oxygen and nitrogen atoms also increases. Analysis of fibres subjected to a range of electrochemical treatments generally shows increased levels of surface oxygen and nitrogen, and surface functional groups such as hydroxyl and carboxyl have also been identified.

The higher modulus, and more highly graphitized fibres, require more intensive treatment conditions to achieve the same level of surface functionality as lower modulus fibres. This is consistent with the view that the graphite basal plane edges, which are more prevalent at the surface of the less highly orientated, lower modulus fibres, are of greater reactivity.

The surface structure of glass fibres is chemically different to organic polymers, and it is not surprising that they do not adhere in a composite. In addition, the silanol groups, present at the surface of amorphous silica based glass fibres, are often hydrated with many monolayers of water. To promote adhesion therefore, a protective size is applied to the fibres, one that contains a coupling agent to bind fibres and matrix together. The initial function of the coupling agent is to render the fibre surface and matrix compatible and aid the wetting process. Chemical bonding between the treated fibre surface and the polymer may then taken place. The effect of the presence of a coupling agent on fibre–matrix adhesion can be illustrated for thermoplastic composites: the transverse flexural strengths of glass reinforced PEEK composites prepared with and without prior application of a coupling agent to the glass fibres, are $127 \, MN/m^2$ and $10 \, MN/m^2$ respectively. Peacock[123] gives a description of some of the strategies that have been followed to optimize the interface in glass fibre reinforced composites. The most successful approaches have been through silane chemistry. Even these have weaknesses. There are difficulties in making a sufficiently stable silane compound which does not degrade at processing temperatures of 350–400°C. Further, the siloxane bond is susceptible to hydrolysis, and conditioning of the composite in water, a standard aerospace test, often reduces fibre–matrix adhesion, with resulting loss of properties. In some cases this effect is reversible, and properties are regained on drying. Much work has been carried out to reduce this problem, including incorporation of hydrophobic groups into the interphase and increasing the density of glass-coupling agent siloxane links. Much work remains to be done.

Achieving good wetting of the large surface area of the reinforcing fibres by highly viscous thermoplastic polymer melts is not a simple matter. To overcome this problem, recourse is often made to a method of presizing the fibres to facilitate this step. Unfortunately, because of the high temperatures required in thermoplastic prepregging, most commercial sizes tend to degrade. Optimization of such sizes is clearly highly desirable, and McMahon[124] speculates that they are essential to the preparation of successful thermoplastic composites. Lind and Coffey[125] and Phillips and Murphy[126] have preferred to presize the fibres with a dilute polymer solution of the matrix polymer prior to melt impregnating by film stacking technology. When the matrix polymer is not conveniently rendered into solution, such as the semi-crystalline polymer PEEK, the expedient of presizing with an amorphous polymer can be used, and Hartness[127] notes success with an obsolete commercial size system to achieve good properties from film stacked PEEK. The use of a size system necessarily defines an 'interphase' between the reinforcement and the matrix.

2.3.2 Chemical bonding

The additional surface chemistry imparted by the surface treatment process renders the fibre more able to bond chemically with the matrix, either during or after the initial wetting process. Chemical bonding mechanisms, such as covalent, polar and donor-acceptor bonds, have all been proposed, and in some cases observed at the fibre–matrix interface[128]. Such interactions may form a significant part of the final adhesive strength of the interface. A notable example where chemistry has been used to advantage in tailoring the interface is the use of alkaline organic salts. These can cause chain extension and even crosslinking in resins such as polyetheretherketone. Such salts can be incorporated on to the surface of carbon fibres, which can cause useful chain extension in PEEK, converting a low molecular weight polymer used to wet the fibres at the impregnation stage into a high molecular weight polymer of excellent toughness[129]. In this case it is the polymer closest to the carbon fibre that is preferentially chain extended, creating a natural adventitious interphase in the material. This approach to active chemistry as a way of creating an interphase in thermoplastic composite materials is most happily exploited when the reaction from the fibre surface is with the end group of the polymer chain, thereby preventing crosslinking reactions that would detract from the thermoplastic character of the system.

2.3.3 Mechanical interlocking

High strength carbon fibres are usually supplied as tows of 3,000, 6,000 or 12,000 filaments wound on to a spool where the tow length is typically 5 km. The fibres are collimated but slightly twisted as a result of the winding process. Inevitably some fibres will be more twisted than others, slightly shortening their length, so that when the tow is placed under tension, these fibres will be subject to a higher stress than their neighbours. In some cases the fibres are deliberately twisted in order to give special effects that may be desirable in weaving. In other cases fibres are entangled, and this can help to open up the fibre tow to penetration by low viscosity polymer solutions. Whenever high viscosity matrix systems are to be impregnated, work must be supplied in order to achieve fibre wetting, and entangled or twisted fibres can be subject to abrasion: with such systems prepreggers usually prefer to work with well-collimated, untwisted fibres.

Hercules AS4 and Courtaulds XAS-0 fibres are circular in cross section, but have slight crenelletions. These are more obvious in the latter fibre. Such crenellations are typically 0.2 µm deep. Other high strength fibres, such as T300 (from Toray and Amoco), can have an irregular shape, often of the form of a kidney. Such shaping has no obvious deleterious effects on fibre packing or composite strength, and may well provide physical keys into which the matrix can lock. I have been unable to find any detailed studies of the effects of fibre geometry on composite performance. At a molecular scale of course all surfaces are rough, and any fibre surface treatment process will affect surface porosity and topography at the molecular level.

Mechanical interlocking alone is unlikely to result in a strong bond when the interface is placed in tension, but will provide a grip when fibre and matrix are placed in shear. A radial and axial compressive grip on a single fibre surrounded by resin can also occur as a result of resin shrinkage during solidification. This effect may be somewhat negated in a high volume fraction composite with closely packed fibres. In the composite the difference in thermal expansion between resin and fibre will cause the resin to shrink on to the fibre in a resin rich region, and away from the fibre in a region where the resin is trapped[130]. Thus the status of stress at the fibre–matrix interface will vary from point to point, dependent on the fibre distribution.

Extensive orientation of polymer molecules in the matrix phase as a result of flow processes is usually only a small and, in some ways undesirable, feature; the matrix polymer may become locally orientated as a result of the constraint supplied by the reinforcing fibres. During solidification of thermoplastic polymers, or during cure of thermosetting systems, the matrix shrinks more than the fibres. The restraint the fibres supply can lead to significant local stresses being induced. Because the fibre reinforcement, which inhibits the matrix shrinkage, is highly anisotropic, the effects of the constraint will also be anistropic inducing a small level of local order in the matrix molecules, most particularly in semi-crystalline polymers.

2.3.4 Crystalline interactions

We have already noted how the crystallization process in certain resins can be nucleated. Nucleation agents can include the surfaces of fibres and also stress concentrations. The organization of the families of spherulites in composite materials can be, in part, determined by the organization of the fibres. In particular, nucleation may take place at the fibre surfaces or, more particularly, where two fibres approach one another.

At 60% by volume of 7 μm fibres, the mean thickness of the resin layer covering each fibre is 1 μm, and workers who have studied interphase effects with single fibres embedded in matrices observe an influence of the fibre surface extending 10 μm or even 100 μm into the resin[131]. It is thus reasonable to assume that the whole matrix phase can be influenced by the fibre surface.

In a simple carbon fibre reinforced semi-crystalline thermoplastic, for example, the fibre surface may influence the nucleation and growth of spherulites within the matrix. An example of this is found in carbon fibre/PEEK composites[132], as illustrated in Figure 2.11.

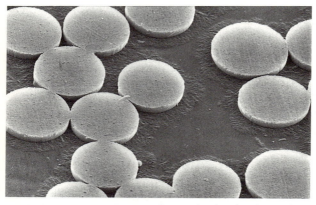

Figure 2.11 Scanning electron micrograph, showing the crystalline texture of partially crystallized PEEK matrix reinforced with high modulus carbon fibre

Nucleation of the spherulites from the surfaces of the high modulus carbon fibres is particularly intense[133,134]. The spherulites very quickly impinge on each other and appear to emerge radially from the fibre surface. This situation has also been observed in other fibre reinforced semi-crystalline thermoplastics, such as nylon and PPS, and has been related to the more graphitic nature of the high modulus

fibre surface rather than to differences in the fibre surface (
graphite plates on the surface of the fibre have a form on to
structure of PEEK can be conveniently superposed (Figure

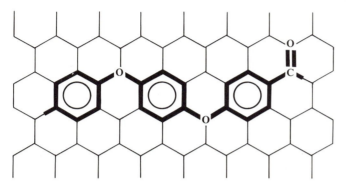

Figure 2.12 Molecular structure of PEEK superposed on a graphite plate

   This superposition is only an approximation to the truth, since it is noted[137] that
the plane of the phenylene rings of PEEK are actually angled with respect to one
another, whereas the graphite plate is flat. Nevertheless this feature can be
assumed to be associated with the ability of graphite to nucleate crystallization in
PEEK. Such nucleation would particularly be expected from high modulus, more
highly graphitized, carbon fibres.
   In the AS4 reinforced PEEK composite on-fibre nucleation is less intense. The
origin of spherulitic growth is almost completely restricted to the point where two
fibres come close together, or where a fibre comes close to another surface. There
will be substantial stresses at the interface because of differences in the thermal
expansion coefficients of fibre and matrix. On-fibre nucleation has also been shown
to be related to shear at the fibre–matrix interface occurring during processing[138].
The effect on composite mechanical properties of on-fibre nucleation is not fully
understood, but it is postulated that a highly orientated layer round the fibres might
improve stiffness in the crystal direction[139]. One supplier[140] recommends
processing PEEK resin composites at very high temperatures in order to eliminate
predetermined nucleii in the matrix phase, thereby favouring crystallization
processes seeded from the fibre surface. Despite this prejudice, there is little direct
evidence that on-fibre nucleation alone is a prerequisite for good fibre–matrix
adhesion; rather it is a consequence of good wetting.

2.4 Thermoplastic structural composite materials

Now that we have established the ingredients of our materials it is time to compare
their performance. Table 2.9 compares the flexural performance of various
structural composites based on polyetheretherketone resin containing 60% by

.ne of fibres. All these materials have been produced by ICI Fiberite, using the .ne melt impregnation technology.

This comparison is made in respect of the flexural properties. It is in flexure that it is easiest, qualitatively, to judge the stiffness of such materials. Flexure is actually a complex deformation, involving tension, compression and shear: in later chapters we shall explore those more fundamental deformations in detail. Of particular significance in this table are the high values of transverse and interlaminar shear properties in comparison with the tensile strength of the resin ($95\,GN/m^2$). This confirms the excellent adhesion between the matrix and the reinforcement. The low values observed for the pitch based sample have two explanations: in the case of the transverse test it is the fibres, not the interface between matrix and fibre, which fail[141]; the so called interlaminar shear or 'short beam shear test' actually leads to a compression failure with these fibres. The modulus and strength data indicate that typically 90% or more of the inherent stiffness and strength of the fibre are realized in these composites.

Table 2.9 Comparative performance of uniaxial structural composites based on PEEK resin at about 60 per cent by volume fibre

Fibre	Designation	Axial flexural modulus GN/m^2	Axial flexural strength MN/m^2	Transverse flexural strength MN/m^2	Short beam shear strength* MN/m^2
High stength carbon	APC-2/AS4	121	1,880	137	105
Intermediate modulus carbon	APC-2/D/IM8	176	1,969	166	112
Pitch based carbon	P75/PEEK	278	728	52†	52
S-glass	APC-2/S-2 glass	54	1,551	157	90
Alumina	APC-2/D/Alumina	120	1,516	186	90

* The test does not provide a clear interlaminar shear failure for these tough matrix composites and compression failure is especially marked for the P75/PEEK sample.
† Low result due to fibre failure.

In Table 2.9 the high strength carbon fibre, AS4, is the standard grade. The intermediate modulus fibres have recently been extensively evaluated for stiffness critical applications. Pitch based, P75, fibres are usually preferred for applications requiring the ultimate in dimensional tolerance, in particular satellite structures. Pitch fibre composites are also used where high thermal conductivity is required. S-glass is the preferred glass fibre used with thermoplastic resins. As well as being used in areas where radar transparency is required, glass fibre composites are also used as an electrically insulating barrier between carbon fibre composite and aluminium structures in order to avoid galvanic corrosion. The ceramic fibres, of which alumina is here the representative, combine a low dielectric constant with high stiffness: their current high price limits them to specialized applications.

The second major variable is the resin phase. There are two significant choices. The first is between semi-crystalline or amorphous polymer, the former being preferred where solvent resistance is a critical factor of service performance. The second choice is the glass transition temperature (Tg), which effectively limits the upper service temperature for structural applications, although it should be noted that semi-crystalline polymers routinely give useful service performance at higher temperatures. These choices are summarized in Table 2.10.

Table 2.10 Thermoplastic matrices for composites

Resin Tg °C	Semi-crystalline matrix	Amorphous	Composite supplier
260		Victrex HTA	ICI Fiberite
260		Avimid K III*	DuPont
260		Torlon C*	Amoco
250		Radel C	Amoco
230		Victrex ITA	ICI Fiberite
230		Victrex PES	Specmat, BASF
220		Ultem PEI	American Cyanamid, Ten Cate
220		Ryton PAS-2	Phillips
220		Radel X	Amoco
205	Victrex HTX		ICI Fiberite
165	Victrex ITX		ICI Fiberite
145		J. Polymer	DuPont
140	Victrex PEEK		ICI Fiberite, BASF
95	Ryton PPS		Phillips

* Avimid K III and Torlon C may be more properly described as linear chain thermosetting polymers, in that they are preimpregnated as prepolymers and then polymerized during processing into the final component. All the other systems noted are fully polymerized, true thermoplastic systems capable of repeated processing.

Several other polymers, particularly polyetherketone variants such as PEK, PEKK and PEKEKK, having intermediate properties between PEEK and HTX, have been proposed. In addition to these continuous fibre reinforced structural composite materials, preimpregnated fibre reinforced products have been prepared on the basis of less stiff resins, including polycarbonate, nylon and polypropylene. Typical of these materials, designed for general industrial use, is the 'Plytron' development product range of materials from ICI. Where, only eight years ago, preimpregnated continuous fibre reinforced thermoplastics was an empty field, today we can select from within the entire range of this versatile family of resins.

The definition of reinforcement and matrix is only the beginning of the story. The quality of the composite material depends critically on the interface between these components, and this, in turn, depends on the method by which those components are assembled. It is this technology of impregnating resin with fibre that we must investigate in Chapter 3.

References

2–1   A. R. BUNSELL, *Fibre Reinforcements for Composite Materials*, Composite Materials Series, **2**, Elsevier (1989).

2–2   W. WATT AND B. V. PEROV, 'Strong Fibres', in *Handbook of Composites*, **1**, Elsevier (1985).

2–3   J. A. PEACOCK AND F. N. COGSWELL, 'Fibre Reinforced Advanced Structural Composites', in I. Miles and S. Rostami, *Multiphase Polymer Systems*, Longman (in press).

2–4   N. J. JOHNSTON AND P. M. HERGENROTHER, 'High Performance Thermoplastics: A Review of Neat Resin and Composite Properties', 32nd International SAMPE Symposium (1987).

2–5   G. H. HARDESTY, 'Poly (Amide-Imide)/Graphite Advanced Composites', Aerospace Congress and Exposition, Longbeach (1984).

2–6   H. H. GIBBS, 'K-Polymer; a New Experimental Thermoplastic Matrix Resin for Advanced Structural Aerospace Composites', 29th National SAMPE Symposium, Reno (1984).

2–7   H. H. GIBBS, 'Processing Studies on K-Polymer Composite Materials', 30th National SAMPE Symposium, pp. 1585–1601 (1985).

2–8   'Torlon' trade literature, TAT-24, Amoco Chemicals Corporation (1984).

2–9   See 2–6.

2–10 G. LUBIN AND S. J. DASTIN, 'Aerospace Applications of Composites', in *Handbook of Composites*, edited by G Lubin, p. 740, Van Nostrand Reinhold (1982).

2–11 T. PEIJS (University of Eindhoven), private communication (1989).

2–12 L. R. SCHMIDT, E. M. LOUGREN AND P. G. MEUSNER, 'Continuous Melt Polymerization of Poly Etherimides', *Int. Polym Processing*, **4**, 4, pp. 270–276 (1989).

2–13 F. N. COGSWELL, D. J. HEZZELL AND P. J. WILLIAMS, 'Fibre-Reinforced Compositions and Methods for Producing Such Compositions', USP4, 559, 262 (1981).

2–14 M. V. WARD, E. NIELD AND P. A. STANILAND, 'Method of Increasing Molecular Weight of Poly(aryl ethers)', EP 0125816B1 (1987).

2–15 W. HILAKOS AND D. J. PATTERSON, 'Poltrusion Apparatus and Method for Impregnating Continuous Lengths of Multifilament and Multifibre Structures', EP 032 654 A2 (1989).

2–16 H. M. COLQUHOUN, C. C. DUDMAN, M. THOMAS, C. A. MAHONEY AND D. J. WILLIAMS, 'Synthesis, Structure and Ring-Opening Polymerization of Strained Macrocyclic Biaryls: A New Route to High Performance Materials' (in press, 1990).

2–17 C.-C. M. MA, H.-C. HSIA, W.-L. LIU AND J.-T. HU, 'Thermal and Rheological Properties of Poly (Phenylene Sulphide) and Poly (Ether Etherketone) Resins and Composites', *Polym Comp*, **8**, 4, pp. 256–264 (1987).

2–18 N. TURTON AND J. McAINSH, 'Thermoplastic Compositions', US Patent 3 785 916 (1974).

2–19 N. J. JOHNSTON, T. K. O'BRIEN, D. H. MORRIS AND R. A. SIMONDS, 'Interlaminar Fracture Toughness of Composites II: Refinement of the Edge Delamination Test and Application to Thermoplastics', 20th National SAMPE Symposium, pp. 502–517 (1983).

2–20 K. E. GOODMAN AND A. R. LOOS, 'Thermoplastic Prepreg Manufacture', *Proc. Am. Soc. for Composite Materials*, 4th Tech. Conf., pp. 746–754, Technomic (1989).

2–21 G. M. WU AND J. M. SCHULTZ, 'Solution Impregnation of Carbon Fiber Reinforced Poly (ethersulphone) Composites', SPE Antec. (1990).

2–22 D. C. LEACH, F. N. COGSWELL AND E. NIELD, 'High Temperature Performance of Thermoplastic Aromatic Polymer Composites', 31st National SAMPE Symposium, pp. 434–448 (1986).

2–23 M COX, 'Liquid Crystal Polymers', *Rapra Review*, **1**, 2, 4 (1987).

2–24 D. S. BAILEY, F. N. COGSWELL AND B. P. GRIFFIN, 'Shaped Articles Formed from Polymers Capable of Exhibiting Anisotropic Melts', EP 044147 (1982).

2–25 F. W. HWANG, D. R. WIFF, C. L. BENNER AND T. E. HELMINIAK, 'Composites on a Molecular Level: Phase Relationships, Processing and Properties', *J. Macromol. Sci. B*, **22**, 2, pp. 231–257 (1988).

2–26 J. M. SCHULTZ, 'Semicrystalline Thermoplastic Matrix Composites', in *Thermoplastic Composite Materials*, edited by L. A. Carlsson, Elsevier Scientific (1991).

2–27 D. DUTTA, H. FRUITAWALA, A. KOHLI AND R. A. WEISS, 'Polymer Blends Containing Liquid Crystals: A Review', *Polym. Engng. and Sci.* (in press, 1990).

2–28 See 2–11.

2–29 M. S. SEFTON, P. T. MCGRAIL, J. A. PEACOCK, S. P. WILKINSON, R. A. CRICK, M. DAVIES AND G. ALMEN, 'Semi-Interpenetrating Polymer Networks as a Route to Toughening of Epoxy Resin Matrix Composites', 19th Int. SAMPE Technical Conf., **19**, pp. 700–710 (1987).

2–30 J. B. ROSE, 'Discovery and Development of the 'Victrex' Polyethersulphones' and 'Discovery and Development of the 'Victrex' Polyaryletherketone PEEK', in *High Performance Polymers. Their origin and Development*, edited by R. B. Seymour and G. S. Kirshenbaum, Elsevier, pp. 169–194 (1986).

2–31 G. R. BELBIN AND P. A. STANILAND, 'Advanced Thermoplastics and their Composites', *Phil. Trans. R. Soc. Lond.*, A 322, pp. 451–464 (1987).

2–32 See 2–15.

2–33 See 2–29.

2–34   P. C. DAWSON AND D. J. BLUNDELL, 'X-ray Data for Poly(aryletherketones)', *Polymer*, **21**, p. 577 (1980).

2–35   P. A. STANILAND, 'Poly(etherketones)', in *Comprehensive Polymer Science*, edited by G. Allen and J. C. Bevington, **5**, pp. 483–497 Pergamon Press (1989).

2–36   J. DEVAUX, D. DELIMOY, D. DAOUST, R. LEGRAS, J. P. MERCIER, C. STRAZIELLE AND E NIELD, 'On the Molecular Weight Determination of a Poly (aryl-ether-ether-ketone) (PEEK)', *Polymer*, **26**, pp. 1994–2000 (1985).

2–37   JOHNSTON AND HERGENROTHER, *op. cit.*

2–38   D. J. BLUNDELL, J. M. CHALMERS, M. W. MACKENZIE AND W. F. GASKIN, 'Crystalline Morphology of the Matrix of PEEK–Carbon Fibre Aromatic Polymer Composites, Part 1: Assessment of Crystallinity', *SAMPE Quarterly*, **16**, 4, pp. 22–30 (1985).

2–39   *Ibid.*

2–40   *Ibid.*

2–41   See 2–34.

2–42   M. A. KING, D. J. BLUNDELL, J. HOWARD, E. A. COLBOURN AND J. KENDRICK, 'Modelling Studies of Crystalline PEEK', *Molecular Simulation*, **4**, pp. 3–13 (1989).

2–43   D. J. KEMMISH, 'Poly(aryl-ether-ether-ketone)', *Rapra Review*, Report 2 (1989).

2–44   See 2–42.

2–45   D. J. BLUNDELL, private communication (1989).

2–46   *Ibid.*

2–47   P. CEBE, private communication (1989).

2–48   See 2–35.

2–49   See 2–38.

2–50   D. J. BLUNDELL AND B. N. OSBORN, 'Crystalline Morphology of the Matrix of PEEK–Carbon Fibre Aromatic Polymer Composites, Part 2: Crystallization Behaviour', *SAMPE Quarterly*, **17**, 1, pp. 1–6 (1985).

2–51   H. X. NGUYEN AND H. ISHIDA, *'Molecular Analysis of the Melting and Crystallization Behaviour of Poly (aryl-ether-ether-ketone)'*, Case Western Reserve University (1985).

2–52   C. N. VELISARIS AND J. C. SEFERIS, 'Crystallization Kinetics of Polyetheretherketone (PEEK) Matrices', *Polymer Engineering and Science*, **26**, pp. 1574–1581 (1986).

2–53   Y. LEE AND R. S. PORTER, 'Crystallization of PEEK in Carbon Fibre Composites', *Polymer Engineering and Science*, **26**, 9, pp. 633–639 (1986).

2–54   D. J. BLUNDELL AND F. M. WILLMOUTH, 'Crystalline Morphology of the Matrix of PEEK-Carbon Fibre Aromatic Polymer Composites, Part 3: Prediction of Cooling Rates During Processing', *SAMPE Quarterly*, **17**, 2, pp. 50–58 (1986).

2–55   P. T. CURTIS, P. DAVIES, I. K. PARTRIDGE AND J. P. SAINTY, 'Cooling Rate Effects in PEEK and Carbon Fibre-PEEK Composites', *Proc. ICCM VI and ECCM 2*, **4**, pp. 401–412 (1987).

2–56   J. HAY AND D. J. KEMMISH, 'Crystallization of PEEK', PRI Polymers for Composites Conference, Paper 4 (1987).

2–57   G. M. K. OSTBERG AND J. C. SEFERIS, 'Annealing Effects on the Crystallinity of Polyetheretherketone (PEEK) and its Carbon Composites', *J. Appl. Polym. Sci.*, **33**, 29 (1987).

2–58   P. CEBE, L. LOWRY, S. Y. CHUNG, A. YAVROUIAN AND A. GUPTA, 'Wide-Angle X-Ray Scattering Study of Heat-Treated PEEK and PEEK Composite', *J. Appl. Polym. Sci.*, **34**, pp. 2273–2283 (1987).

2–59   J. F. CARPENTER, 'Thermal Analysis and Crystallization Kinetics of High Temperature Thermoplastics, *SAMPLE J.*, **24**, 1, pp. 36–39 (1988).

2–60   D. J. BLUNDELL, R. A. CRICK, B. FIFE AND J. A. PEACOCK, 'The Spherulitic Morphology of the Matrix of Thermoplastic PEEK/Carbon Fibre Polymer Composites', in *New Materials and their Applications, 1987*, edited by S. G. Burney, Institute of Physics (1988).

2–61   M.-F. SHEU, J.-H. LIN, W.-L. CHUNG AND C.-L. ONG, 'The Measurement of Crystallinity in Advanced Thermoplastics', 33rd International SAMPE Symposium, pp. 1307–1318 (1988).

2–62   D. E. SPAHR AND J. M. SCHULTZ, 'Determination of the Matrix Crystallinity of Composites by X-ray Diffraction', submitted to *Polymer Composites* (1988).

2–63   P. CEBE, 'Application of the Parallel Avrami Model to Crystallization in PEEK', *Polymer Engineering and Science*, **28**, 18, pp. 1192–1197 (1988).

2–64   M. QI, X. XU, J. ZHENG, W. WANG AND Z. QI, 'Isothermal Crystallization Behaviour of Poly(ether-ether-ketone) (PEEK) and its Carbon Fibre Composites', *Thermochimica. Acta*, **134**, pp. 223–230 (1988).

2–65   P. CEBE, L. LOWRY AND S. CHUNG, 'Use of Scattering Methods for Characterization of Morphology in Semi-Crystalline Thermoplastics', SPE ANTEC (1989).

2–66   M. DAY, T. SUPRUNCHUK, Y. DESLANDES AND D. M. WILES, 'The Thermal Processing of Poly (Ether-Ether-Ketone) (PEEK): Some of the Factors that Influence the Crystallization Behaviour', 34th International SAMPE Symposium, pp. 1474–1485 (1989).

2–67   D. J. BLUNDELL, R. A. CRICK, B. FIFE, J. A. PEACOCK, A. KELLER AND A. J. WADDON, 'The Spherulitic Morphology of the Matrix of Thermoplastic PEEK/Carbon Fibre Aromatic Polymer Composites', SPE 47th ANTEC, pp. 1419–1421 (1989).

2–68   D. J. BLUNDELL, R. A. CRICK, B. FIFE, J. A. PEACOCK, A. KELLER AND A. WADDON, 'Spherulitic Morphology of the Matrix of Thermoplastic PEEK/Carbon Fibre Aromatic Polymer Composites', *Journal of Materials Science,* **24**, pp. 2057–2064 (1989).

2–69   G. M. K. OSTBERG AND J. C. SEFERIS, 'Annealing Effects on the Crystallinity of Polyetheretherketone (PEEK) and Its Carbon Fiber Composite', *Journal of Applied Polymer Science,* **33**, pp. 29–39 (1987).

2–70   H. MOTZ AND J. M. SCHULTZ, 'The Solidification of PEEK. Part I: Morphology', submitted to *Journal of Thermoplastic Composite Materials* (1989).

2–71   H. MOTZ AND J. M. SCHULTZ, 'The Solidification of PEEK. Part II: Kinetics', submitted to *Journal of Thermoplastic Composite Materials* (1989).

2–72   F. MEDELLIN-RODRIGUEZ AND P. J. PHILLIPS, 'Crystallization Studies of PEEK Resin', SPE ANTEC '90, pp. 1264–1267 (1990).

2–73   See 2–34.

2–74   See 2–72.

2–75   A. J. LOVINGER AND D. D. DAVIS, 'Single Crystals of Poly (ether-ether-ketone) (PEEK)', *Polymer Communications,* **26**, pp. 322–324 (1986).

2–76   A. J. LOVINGER AND D. D. DAVIS, 'Solution Crystallization of (Poly ether ketone)', *Macromolecules,* **19**, pp. 1861–1867 (1986).

2–77   A. J. WADDEN, M. J. HILL, A. KELLER AND D. J. BLUNDELL, 'On the Crystal Texture of Linear Polyarols (PEEK, PEK and PPS), *J. Materials Science,* **22**, pp. 1773–1784 (1987).

2–78   See 2–68.

2–79   See 2–43.

2–80   H. VOSS AND K. FRIEDRICH, 'On the Wear Behaviour of Short Fibre Reinforced PEEK Composites', *Wear,* **116**, pp. 1–18 (1987).

2–81   T. SASUGA, N. HAYAKAWA, K. YOSHIDA AND M. HAGIWARA, 'Degradation in Tensile Properties of Aromatic Polymers by Electron Beam Irradiation, *Polymer,* **26**, pp. 1039–1045 (1985).

2–82   See 2–43.

2–83   See 2–1.

2–84   See 2–2.

2–85   P. G. RIEWALD, A. K. DHINGRA AND T. A. CHAN, 'Recent Advances in Aramid Fibre and Composite Technology', Sixth International Conference on Composite Materials, **5**, p. 362 (1987).

2–86   *High Tech Materials Alert,* Technical Insights Incorporated, Jan. 2 (1989).

2–87   D. HULL, 'An Introduction to Composite Materials', Cambridge University Press (1981).

2–88   S. KHAN, J. F. PRATTE, I. Y. CHANG AND W. H. KRUEGER, 'Composite for Aerospace Application from Kevlar Aramid Reinforced PEKK Thermoplastic', 35th International SAMPE Symposium, pp. 1579–1595 (1990).

2–89   R. M. TURNER AND F. N. COGSWELL, 'The Effect of Fibre Characteristics on the Morphology and Performance of Semi-Crystalline Thermoplastic Composites', *SAMPE J.,* **23**, 1, pp. 40–44 (1987).

2–90   W. P. HOOGSTEDEN AND D. R. HARTMAN, 'Durability and Damage Tolerance of S-2 Glass/PEEK Composites', 35th International SAMPE Symposium, pp. 1118–1130 (1990).

2–91   J. E. GORDON, *The New Science of Strong Materials,* Penguin Books (1976).

2–92   F. E. WAWNER, 'Boron Filaments', in L. Bortman and R. Kvock (eds.), *Modern Composite Materials,* Addison-Wesley Pub. Co., Reading, Mass. (1967).

2–93   U. MEASURIA, 'A New Fibre Reinforced Thermoplastic Composite for Potential Radome Application: PEEK/Alumina', 3rd International Composites Conference, Liverpool 23–25 March (1988).

2–94   N. YAJIMA, 'Silicon Carbide Fibres', in *Handbook of Composites I. Strong Fibres,* edited by W. Watt and B. V. Peron, Elsevier (1985).

2–95   P. BRACKE, H SCHURMANS AND J. VERHOEST, *Inorganic Fibres and Composite Materials',* EPO Applied Technology Series, Vol 3, Pergamon Int. Infor. Corporation (1984).

2–96   A. SHINDO, 'Studies on Graphite Fibre', Report 317, Government Ind. Res. Inst., Osaka, Japan (1961).

2–97   W. WATT, L. N. PHILLIPS AND W. JOHNSON, *The Engineer,* London, 221, 815 (1966).

2–98   See 2–96.

2–99   See 2–2.

2–100 J. B. DONNET AND R. G. BANSAL, *Carbon Fibres,* International Fibre Science and Technology, Marcel Dekker Inc. NY (1984).

2–101 A. OBERLIN AND M. GUIGON, 'The Structure of Carbon Fibre', in *Fibre Reinforcements for Composite Materials',* Elsevier (1989).

2–102 M. S. DRESSELHAUS, G. DRESSELHAUS, K. SUGIHARA, I. L. SPAIN AND H. A. GOLDBERS, *Graphite Fibres and Filaments,* Springer Series in Materials Science, **5** (1988).

2–103 D. J. JOHNSON, 'Structure and Properties of Carbon Fibres', in *Carbon Fibres, Filaments and Composites,* edited by J. L. Figueuredo, C. A. Barnado, R. T. K. Baker and K. J. Huttinger, NATO ASI Series, **177,** pp. 119–146, Kluwer Academic (1990).

2–104 O. L. BLAKSLEE, D. G. PROCTOR, E. J. SELDIN, G. B. SPENCE AND T. WENG, *J. Appli. Phys.,* **41,** 8, p. 3373 (1970).

2–105 See 2–101.

2–106 M. ENDO, 'Structure of Mesophase Pitch-based Carbon Fibres', *J. Mat. Sci.,* **23,** 598–605 (1988).

2–107 See 2–87.

2–108 See 2–100.

2–109 K. F. ROGERS, L. N. PHILLIPS, D. M. KINGSTON-LEE, B. YATES, M. J. OVERY, J. P. SARGENT AND B. A. McCALLA, 'The Thermal Expansion of Carbon Fibre Reinforced Plastics', Part 1, *J. Materials Science,* **12,** pp. 718–734 (1977).

2–110 G. WAGONER AND R. BACON, 'Elastic Constants and Thermal Expansion Coefficiently of various Carbon Fibres', Penn State Conference, pp. 296–297 (1989).

2–111 P. M. SHEAFFER, 'Transverse Thermal Expansion of Carbon Fibres', *Proc. XVIIIth Biennial Conference on Carbon,* pp. 20–21 (1987).

2–112 See 2–109.

2–113 See 2–91.

2–114 See 2–3.

2–115 A. LUSTIGER, 'Morphological Aspects of the Interface in the PEEK–Carbon Fiber System', SPE ANTEC '90, pp. 1271–1274 (1990).

2–116 B. FIFE, J. A. PEACOCK AND C. Y. BARLOW, 'The Role of Fibre-Matrix Adhesion in Continuous Carbon Fibre Reinforced Composites: A Microstructural Study', ICCM-6, **5,** 439–447 (1987).

2–117 J. A. BARNES, 'Thermal Expansion Behaviour of Thermoplastic Composites', submitted to *J. Materials Science* (1990).

2–118 See 2–3.

2–119 D. HODGE, B. A. MIDDLEMISS AND J. A. PEACOCK, 'Correlation Between Fibre Surface Energetics and Fibre Matrix Adhesion in Carbon Fibre Reinforced PEEK Composites', in 'Tailored Interfaces in Composites', edited by G. C. Pantans and E. J. H. Chen, *Materials Research Society Proceedings,* **170,** Boston (1989).

2–120 See 2–3.

2–121 See 2–119.

2–122 See 2–119.

2–123 See 2–3.

2–124 P. E. MCMAHON, 'Thermoplastic Carbon Fibre Composites', in *Developments in Reinforced Plastics – 4,* edited by G. Pritchard, pp. 1–30 (1984).

2–125 D. J. LIND AND V. J. COFFEY, 'A Method of Manufacturing Composite Material', British Patent 1,485,586 (1977).

2–126 L. N. PHILLIPS AND D. J. MURPHY, 'Stiff, Void Free Fibre Reinforced Thermoplastic Polymer Laminate Manufacture', British Patent 1,570,000 (1980).

2–127 J. T. HARTNESS, 'Thermoplastic Powder Technology for Advanced Composite Systems', 33rd International SAMPE Symposium, pp. 1458–1472 (1988).

2–128 See 2–3.

2–129 See 2–13.

2–130 F. N. COGSWELL, 'The Processing Science of Thermoplastic Structural Composites', *International Polymer Processing,* **1,** 4, pp. 157–165 (1987).

2–131 F. N. COGSWELL, 'Microstructure and Properties of Thermoplastic Aromatic Polymer Composites', 28th National SAMPE Symposium, pp. 528–534 (1983).

2–132 J. A. PEACOCK, B. FIFE, E. NIELD AND C. Y. BARLOW, 'A Fibre-matrix Study of some Experimental PEEK-Carbon Fibre Composites, in *Composite Interfaces,* edited by H. Ishida and J. L. Koenig, p. 143, Elsevier (1986).

2–133 See 2–115.

2–134 See 2–132.

2–135 T. BESSELL AND J. B. SHORTALL, 'The Crystallization and Interfacial Bond Strength of Nylon 6 at Carbon and Glass Fibre Surfaces', *Material Science,* **10**, pp. 2035–2043 (1975).

2–136 A. J. WADDON, M. J. HILL, A. KELLER AND D. J. BLUNDELL, 'On the Crystal Texture of Linear Polyaryls', *Material Science,* **22**, pp. 1773–1784 (1987).

2–137 See 2–42.

2–138 M. J. FOLKES' results, presented at the conference 'Flow Processes in Composite Materials', Brunel University (1988).

2–139 J. L. KARDOS, F. S. CHENG AND T. L. TOLBERT, 'Tailoring the Interface in Graphite-Reinforced Polycarbonate', *Polym. Engineering and Science,* **13**, pp. 455–461 (1973).

2–140 A. C. HANDERMANN, 'Advances in Comingled Yarn Technology', 26th Int. SAMPE Technical Conference, pp. 681–688 (1988).

2–141 J. A. BARNES AND F. N. COGSWELL, 'Thermoplastics for Space', *SAMPE Quarterly,* **20**, 3, pp. 22–27 (1989).

3 Product forms

'Wherefore by their fruits ye shall know them.' Matthew vii, 20

Having identified the ingredients of thermoplastic composites, we must next investigate how they can be assembled into material forms suitable for configuration into structures. In turn this will take us to a study of the forming processes and the properties of the optimum microstructure. A common experience of polymer science is that there is an inverse relation between ease of processing and service performance[1]. It must be emphasized that this is not a scientific law, only a reflection of empiricism; nevertheless it warns us that some compromise may have to be sought in developing an 'optimized' product form, or that we may need more than one product form of a material in order to achieve all the aspects of shaping and performance we require for a particular application. Remembering that the interface between matrix and reinforcement is critical to the quality of a composite material, we can identify two distinct strategies. In one strategy the resin is incorporated into the fibres and the interface is established in a preimpregnation process, the product of which is often called a 'prepreg'. The second major strategy is to assemble the matrix and reinforcement together, for example as mixtures of fibres, and to make the interface between those components at the same time as the structure is formed. There are a host of secondary strategies to select from: collimated fibres or woven fabrics; continuous or long discontinuous fibres; straight or crimped fibres; simple or sophisticated systems, for example with an additional fugitive component added either adventitiously or by design; and the use of chain extension after shaping. Thermoplastic composites, and especially carbon fibre reinforced polyetheretherketone, are an especially rich field for product forms. That diversity of form can be a boon to the informed designer, who can select to meet his requirements, but it has also led to some confusion. All carbon fibre reinforced polyetheretherketone composites are not equal.

3.1 Impregnation after shaping

The processing technologies that dominate the thermoset composite industry are hand lay up and autoclave cure. These are laborious and lengthy processes but well suited to the manufacture of relatively low numbers of large complex structures.

The aerospace community has a heavy investment in such technology, and has encouraged the development of product forms which can utilize that investment. A rich diversity of solutions has been provided to deliver systems that can meet those requirements. They exploit a protracted moulding cycle in order to form the interface between resin and fibre, and in some cases to form the resin itself.

3.1.1 In-situ polymerization

The principle of resin injection, whereby a monomer is incorporated into a fibrous preform, is usually exploited with crosslinkable thermosetting resins[2,3,4]. Certain monomers and prepolymers can utilize these processes and be subsequently polymerized to linear chain thermoplastic polymers. Such polymerization processes are usually accompanied by significant shrinkage of the resin phase, which may leave voids in the structure. In the case of thermoplastic composites such voids could be subsequently moulded out by a consolidation process above the melting point.

Resin injection has, so far, not been widely exploited with thermoplastic polymers. One possible monomer is methylmethacrylate, and composite structures based on polymethylmethacrylate polymerized *in-situ* have been considered as candidate materials for space structures[5] because of outstanding resistance to ultraviolet light. There has also been interest in the development of ring-opening prepolymers[6] of the class of engineering polymers, and particular emphasis has been laid on their low viscosity and suitability as impregnant systems for subsequent polymerization. A ring-opening route to polyketones has been identified[7].

Resin injection into fibrous preforms has two identified benefits. The preforms are normally made with reinforcement in several directions, and thereby interlaminar weakness can be eliminated. This is a very real advantage with brittle matrix composites but, in the case of tough thermoplastics, the losses in stiffness due to fibre crimping and out of plane fibres may more than offset the additional toughness gain. Real advantages are to be gained in making shapes by exploiting 10,000 years of textile technology, and this potential will ensure that this route is frequently re-examined.

3.1.2 Linear chain thermosetting 'prepregs'

A variant of *in-situ* polymerization is to preimpregnate the fibres with a prepolymer and subsequently shape and cure. This, of course, is the dominant thermoset technology for epoxy resins, and so immediately fulfils the requirements of a system for hand lay up and autoclave processing. This approach allows the 'prepreg' to be made by existing technologies[8].

Although a 'prepreg' is used, this is really a 'prepolymer prepreg', and the interface between matrix and reinforcement cannot be established until after the component is shaped and the matrix formed. Further, a surprising feature of optimized 'prepregging' of thermoset resins is that it is preferred NOT to fully wet out the fibres at the 'prepreg' stage. The unwetted fibres actually present an essential pathway for the extraction of volatile materials from the interior of large, thick mouldings[9], and with condensation reaction polymers, which are the most widely used systems[10,11], there are usually large volumes of volatiles to be extracted. The curing process of such resins as the polyamideimide family is traced by Gibbs[12], who notes that the prepolymer, initially a tacky liquid, first polymerizes to a powdery solid, which facilitates removal of volatiles, before finally melting, fusing and wetting the fibres.

3.1.3 Film stacking

For a long time film stacking was the standard technology for making thermoplastic composites[13]. Layers of reinforcing fibres are laminated between layers of thermoplastic polymer film. This stack is then fused, usually under a pressure of at least 10 atmospheres, for about 1 hour before being solidified. During the protracted high pressure process the resin flows in between the fibres, and the interface is formed. To aid in this stage of the process, it is usually considered desirable to have the fibres presized with a compatible resin[14,15]. The difficulty of wetting out the fibres with matrix has been theoretically studied by Lee and Springer[16], who confirm the experimental finding that several hours are required to fully wet out the fibres. The obvious difficulty of forcing a highly viscous liquid through the micron sized gaps between fibres is overcome by applying pressure but this also forces the fibres together (Figure 3.1), packing them in such a way as to

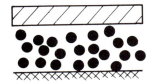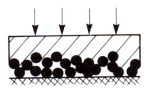

Figure 3.1 Fibre bed consolidation under pressure

make infiltration by the resin more difficult still. A typical stack of reinforcement to be wetted out would be about fifteen fibres deep. To achieve wetting of the reinforcement by the resin under these circumstances requires high work input. Unfortunately that same work input can damage the reinforcing fibres, particularly in the case of woven fabric forms where the fibres cross one another. This damage can be alleviated by the use of protective presizing systems. Even so, the strength properties of film stacked composites fail to achieve full utilization of the inherent fibre properties[17]. The problems of wetting the fibres in a film stacking process are also encountered to a lesser degree with other, more intimately mixed, systems. Since film stacking is based on flat film, it is a technology well adapted to making flat sheet, but only with dexterity can it be adapted to shaping complex forms : this difficulty is compounded by the difficulty of applying high lateral consolidation forces to wet out the non-horizontal sections of a moulding. Despite these manifest constraints, film stacking is still widely used as the most convenient technology for preparing small quantities of bespoke composite. Some of the earliest work on carbon fibre reinforced polyetheretherketone utilized this technology[18].

3.1.4 Partially impregnated tows

A variant on continuous film stacking is to sandwich a layer of reinforcing fibres between two films. The films can be provided by a simple melt coating process. This approach has been particularly favoured for use with long discontinuous fibres[19]. The advantage claimed for such long discontinuous fibres as partially impregnated tows is that they can be stretched, allowing the discontinuous fibres to slip past one another, but, because the fibres are long (typically several centimetres), the mechanical properties are not compromised[20]. This ability to stretch the prepreg is said to be particularly advantageous when forming complex shapes. Unfortunately a natural tendency for such a system is that, as the prepreg stretches, it becomes thinner, and so weaker. If the stretch is non-uniform, it will become unstable, leading to localized rupture. This stability problem would be particularly severe if the long discontinuous fibres were fully wetted by the resin, since the viscous drag between fibres would be exceedingly high. Stability is probably enhanced by carrying out the forming process when the fibrous network is unwetted, and therefore weak in comparison to the layers of polymer film adhering to either side. These films dominate the flow and carry the fibres with them; the same principle is exploited in diaphragm forming[21]. This product form has been demonstrated with a polyetherketone resin[22], and encouraging properties are reported.

3.1.5 Cowoven and commingled products

Film-stacking technology has two identified limitations: it is not readily drapable into form, and the reinforcement and matrix are not integrated with sufficient intimacy. These two constraints can be loosened by using the matrix in fibre or narrow tape form, and coweaving or intermingling it with the reinforcing fibres. This approach was described for PEEK composites by Slater[23] and was developed under NASA funding[24]. It has had many champions[25-31]. Besides flaccid drapability, such fibrous forms can be knitted or braided into complex 3-D structures that can subsequently be consolidated[32-38]. Such 3-D structures have a guaranteed, coarse, distribution of resin in comparison with the option of *in-situ* polymerization (Section 3.1.1 above), but there are significant problems with debulking such structures. Such debulking can induce crimp in the fibres, and failure to remove all the air can cause degradation during the melting process.

The theory of commingled product forms is to place in juxtaposition resin and reinforcement fibres, such that, when melted, the matrix instantly coats each of the fibres. It is certainly possible to define a simple array of fibres where this would be so (Figure 3.2). Unfortunately statistics would have it otherwise, and a typical random packing would be that shown in Figure 3.3, where typically only 50% of the

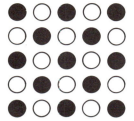

Figure 3.2 Idealized array 50/50 of resin and reinforcement fibres of the same diameter

Figure 3.3 Random packing 50/50

reinforcing fibre surface is intimately contacted by the resin at first melting. This problem is further exacerbated by the difficulty of spinning very fine resin fibres. Carbon fibres usually have diameters of 7 μm, but the normal commercial grades of commingled product contain 40% by volume of resin fibres about 20 μm in diameter. In such circumstances there is four and a half times as much surface area of carbon fibre as there is of resin, so that, even with the most advantageous distribution, the initial wetting can only account for 20% of the total surface area of carbon fibre, leaving a lot of work still to do. Of course the bundles of carbon fibre that remain to be wetted are significantly smaller than is the case with film stacking, and more rapid wetting at lower applied pressure is usually reported for these systems[37]. The quest for fine matrix fibres often directs that, in the spinning process, those fibres are drawn and so oriented. When melted, that orientation can relax, causing the fibres to shrink. If uncontrolled, this shrinkage may cause distortion of the commingled fabric. It follows that it is highly desirable that such fabrics should be constrained while they are melted. Olson[38] reviews manufacturing technologies for working with commingled yarns; he emphasizes the need for high temperatures and pressures and long consolidation times.

One variant on the commingled product form is to spin together stretch broken, long discontinuous fibres of matrix and reinforcement – the Heltra Filmix system[39–42]. This approach necessarily incorporates some twist into the fibres, which makes the tows slightly stretchable. The twist also stabilizes the distribution of resin and reinforcement fibres, which may segregate in simple commingled products. This twist is reported to give improved uniformity of resin dispersion[43–44]. The twist also has a negative effect, in that the orientation of the fibres is slightly off axis, but this is balanced by observations of enhanced interlaminar shear strength[45]. Russell[46] has studied processing techniques for this class of material, and recommends press forming as opposed to autoclave technology in order to achieve adequate consolidation. Because this type of product form is designed to be used in complex geometry components, the loss of stiffness may not be too critical. Morales[47] notes that, although the simple coupon properties of Heltra Filmix are inferior to those of simple commingled products, in complex structures they can deliver enhanced performance.

The ability to knit, braid or weave complex shapes, which are subsequently consolidated, opens up the arsenal of textile technology to the designer. This is an active field at the present time (1990), although there has yet to be conclusive demonstration that any gains in toughness[48–50], in comparison to simple laminated structures, outweigh the loss in stiffness due to crimping of the fibres during debulking of the woven preforms: what is an obvious advantage for brittle matrix systems such as carbon–carbon does not necessarily carry through to inherently tough composites.

3.1.6 Powder coating

It is possible to impregnate the fibre bundles with powder as a precursor to melting in order to wet the fibre with the matrix. Such a technique, whereby opened rovings were passed through a fluidized bed of powder and subsequently melted in a hot tube, was discussed by Price[51], who emphasized that this process was designed to create a partial wetting of the fibres that could be completed in subsequent processing operations. The key problem, as with commingling, is achieving a fine enough dispersion of the matrix. As a precondition, this requires fine matrix powders. The fracture mechanics of grinding[52] state that it is not possible to comminute materials below their critical flaw size, and for tough matrices such as polyetheretherketone this sets a lower limit of about 25 μm. Of course protracted grinding at cryogenic temperatures can further reduce the particle size, but only at significant cost. In the case of PEEK resin it is possible, by careful definition of the crystalline morphology, to create a fragile network that may be ground down to sub-micron particles[53], but fine powders are not conveniently air-fluidized: such fine powders must usually be worked from dispersions. Workers at Georgia Tech[54–57] have developed the fluidized bed coating technique to distribute coarse powder on to a spread tow. Spreading of the tow is a critical stage, and a preferred version of the process uses an air knife. Further spreading is achieved by the use of electrostatics, which can also help to pin the powder to the fibre. This powder is subsequently melted, providing distributed nodules of polymer but allowing the uncoated fibres to make the 'tow preg' flexible, so that it can be easily woven. Muzzy[58] claims good properties and excellent economics for this product and process.

It is not always necessary to fuse the powder to the fibres. Throne and Soh[59] demonstrate the benefits of electrostatic dry powder prepregging when using powders of the order 5–10 μm. A flexible preform, which can be woven and braided, is achieved by first wedging the powder into the fibres and then coating the powder-impregnated tow in a 'wire covering' operation[60]. This technology, termed FIT (Fibre Imprégnée Thermoplastique), has been applied to make carbon fibre reinforced polyetheretherketone, but considerable difficulty seems to have been experienced in achieving a satisfactory interface in consolidated mouldings[61].

Dispersions have been used to make 'powder prepregs' that have tack and drape[62–64]. As well as the reinforcement fibre and resin powder, this system includes a tackyfying agent that sticks the resin particles to the fibre surface. When wetted, the 'prepreg' ply becomes tacky, so that it can be assembled by hand and conformed to shape before drying and consolidating. This system, designed for hand lay up and autoclave technology, was commercialy launched in 1988 but withdrawn for further development a year later. While good properties were claimed for this product form, there was difficulty with debulking in complex curvature structures, and some mystery about what happened to the tackyfying agent during the consolidation process.

Coating of fabrics with PEEK powder was described by Chang[65], but the properties achieved were inferior to film stacked compositions[66], although the product presumably costs less.

Powder materials are attractive as a starting point for composite manufacture. Many high performance polymers are polymerized as powders and, in his armoury, the chemist has technologies, such as emulsion and suspension polymerization, by which he can control particle size. Suspensions, certainly water suspensions, can be handled in a variety of standard chemical engineering equipment. Suspensions in non-aqueous media may be preferred if the need for surfactants is reduced thereby. Of all the methods currently being studied, the patent literature[67–72] suggests that powder, or dispersion technology, is the preferred route for most investigators.

3.1.7 General observations on impregnation after shaping

When seeking to create an interface between matrix and fibre, we must pay close attention to both the physics and the chemistry of that interface. In order to fuse the resin to the surface of the fibre it is necessary to displace any air lying in between. The fusion necessarily takes place at a high temperature, above the melting point of the resin. At such temperatures the air, and more specifically the oxygen, may interact both with the surface of the fibre and with the resin. Since oxidation is a surface treatment that enhances the activity of carbon fibre surfaces further oxidation may alter that surface. It is known that prolonged exposure of fibres to temperatures as low as 300°C[73] can result in detreatment, or loss of functional groups, and so loss of adhesion capability. In respect of the resin matrix, oxidation is a significant concern. Polyetheretherketone will crosslink in the presence of oxygen at high temperature, and in these circumstances this crosslinking will necessarily occur at the free surface that was to be bonded to the fibre. From these arguments we can conclude that, in order to achieve a quality controlled interface, it is desirable to make the bond at the matrix reinforcement interface in the absence of air and as rapidly as possible. This need is the primary motivation of a preimpregnation process.

3.2 Preimpregnation processes

Desirable as a preimpregnation process is, achieving full wetting of all the fibres is not easily done. We have already (Chapter 1) identified that in 10 ml of composite we have nearly 4 m^2 of surface area of fibre to be coated with a thimbleful of matrix resin. If the thimbleful of resin has fluid consistency, then this is no great problem; however, in the case of thermoplastic polymers, the resin, even when molten, has the consistency of well chewed chewing gum, and the problem of spreading that matrix over so large an area is obvious. Further, economics demand that the preimpregnation process should be rapid, so that making our 10 ml of composite (approximately 16 g or 0.5 m of prepreg 150 mm wide) must be achieved in a fraction of a minute.

In the analogy with chewing gum you have less than a minute to spread two pieces of chewing gum over the area of a large dining table, including the time taken to soften the gum. The combination of a desirable end point obstructed by a clear physical process barrier has generated a range of engineering answers. The details of most of these are held as proprietary technology by the 'prepreg' manufacturers. Inspection of the patent literature indicates three basic approaches: solutions, chain extension and melts. Elements of those approaches are often combined within a single process. Until the early 1980s no commercially workable answer to the problem had been found; however, since the introduction of Aromatic Polymer Composite 'prepregs'[74], the existence of a solution had been demonstrated and has stimulated a range of other researchers.

3.2.1 Solution impregnation

The most frequently tried approach to impregnation with polymers is to dissolve them in a suitable solvent and use that low viscosity medium to wet the fibres. Solution impregnation of high performance aromatic polymers such as polyether-sulphone was first demonstrated by Turton and McAinsh[75–76]. They emphasized the requirement of low viscosity, and demonstrated that preferred solution ranges were at about 10% by weight of polymer. This range is confirmed by other workers[77,78]. Such a solution provides a low viscosity medium which can easily wet fibres. It also increases tenfold the volume of the coating medium, and thereby reduces the requirement to flow by such a factor. Our thimble full of chewing toffee has been converted into a wineglass full of olive oil, and the problem of spreading over the area of a dining table is accordingly reduced – indeed the problem now is one of containment. Solution impregnation, especially of fabrics, is a widely used technique with thermosetting resins prepolymers: it is a technology for which much equipment is already available. Such solution technology operates at ambient, or near ambient, temperature, and is followed by conventional drying processes to eliminate the now unwanted solvent. The wetting process itself relies mainly on surface tension forces, and no additional work needs to be supplied. The process is relatively easy to operate, provided that a suitable solvent is available for the matrix polymer.

Despite the powerful arguments in favour of solution technology, and its early demonstration, it was for many years a relatively unused approach. The reasons against solution impregnation, learnt usually after considerable expense of trial and frustration, and relearnt with a depressing frequency, are the following: the use of solution technology usually betrays a sensitivity to solvents in service; the presence of residual solvents compromises processing and reduces service performance, particularly at high temperature; because the fibre wetting process is controlled adventitiously by surface tension forces, the fibre distribution is less under the control of the prepregger; and the interface established in the solution process leaves something to be desired. Each of these problems can be addressed, but a satisfactory solution changes an apparently simple process into a sophisticated operation.

To be processed by solution, the polymer must be soluble. Highly volatile solvents, such as methylene dichloride, are often preferred as being very effective for a wide range of polymers: methylene dichloride is also a component of many commercial paint strippers and cleaning agents, to which composite structures may often be exposed. A popular solvent for prepregging aromatic polymers is the rather less volatile n-methyl pyrrolidone. Mixed solvent systems, tailored most efficiently to dissolve the polymers, also have their champions[79]. Other workers prefer instead to tailor the polymer molecule by the use of copolymer systems in order to increase solubility of the resin[80].

The problem of solvent sensitivity in service can be addressed in one of three ways: the use of the composite can be constrained to those applications where solvent exposure is not critical or, in use, those solvents to which it is susceptible can be eliminated; the linear chain polymer can, after the composite is formed, be converted into a non solvent sensitive system by crosslinking; or the 'solvent' used can itself be a monomer and be subsequently polymerized. The option of constraining the application of the composite to areas of non-exposure to solvents requires one of two scenarios: either the structure is of such a valuable and critical kind that the reagents to which it is exposed can be carefully monitored and controlled, or the structure is sufficiently limited in service requirements that accidental exposure to an unkind solvent would not critically limit its function. The option of carrying out further chemistry on the linear chain polymer, or prepolymer, has been discussed before (Chapter 2), but this may constrain the thermoplasticity of the composite. One elegant solution is to use a monomer, or prepolymer, as a solvent for the thermoplastic chain, and subsequently polymerize that 'solvent', usually in a crosslinking reaction. This is preferably done in such a way that the thermoplastic chain is incorporated into that reaction, forming thereby an interpenetrating network: this approach is used by McGrail[81] to form the toughest thermosetting compounds available today. Despite the solubility of amorphous thermoplastic polymers, they are increasingly finding use today in non-critical applications and in application areas where other properties that they possess, such as high temperature performance, are deemed to be so important that their exposure to chemical reagents can be controlled.

Elimination of solvents from the composite takes place in two stages. The removal of the bulk of the solvent – increasing the polymer concentration from about 15% to 95% – usually represents a major, even the major, element of the prepregging operation. Although this is usually seen to be conventional chemical engineering, extraction and reclaim of large quantities of solvent (typically 3 kg of solvent for every 1 kg of prepreg made) can be an expensive operation. The second stage, removing the final traces of solvent, is particularly difficult. The presence of residual solvent may, on occasion, be exploited. It can be used to soften the prepreg to make it drapable and tacky, and therefore suitable for hand lay up technologies: in this case the residual solvent must be removed during the processing to end product stage. Failure to remove the residual solvent leads to bubbles or blisters in the composite, and the high temperature performance can also be compromised[82,83]. The presence of even traces of residual solvent may also provide a pathway for accelerated attack by solvents in service. The difficulty of eliminating such residual solvent is testified to by Goodman and Loos[84]. Today it is generally recognized that this cannot be achieved without heating the polymer to the melt and preferably applying vacuum. Despite these difficulties, a range of solution impregnated thermoplastic composites is available today, and some of them are described as solvent free[85–87].

Quality in a prepreg requires control of the resin fibre distribution. Appendix 7 addresses some of the control issues encountered during solution impregnation, and more specifically during solvent removal. These include issues of resin–fibre distribution and the interface. It requires considerable attention to detail and process optimization to make high quality prepregs by the solution route[88].

We have identified the interface between reinforcement and matrix as a critical component of thermoplastic composite materials. There is evidence[89,90] that the interface achieved in solution impregnated materials is less than optimum. The reasons for this are, so far, only the subject of speculation. Such speculations fall into three groups: physical detachment of the interface during the solvent extraction stage, a concentration of residual solvent at the interface, and a phase separation of low molecular weight species at the surface of the fibre. During removal of the solvent from the interior of the composite the natural course of escape is through the surface skins. An alternative course is for the solvent to move to the resin–fibre interface, detach that interface and then wick along the fibre surface until the twisting of the fibres causes them to approach the surface, where a sufficient accumulation of solvent can literally explode through to make its escape. Direct evidence of this argument is hard to come by, but snapping a piece of solution impregnated prepreg frequently reveals bare fibres, which certainly should have been wetted by the polymer solution at the impregnation stage. Micrographs of solution prepregged materials (Figure 3.4), sometimes give evidence of detachment of the resin from the fibre surface. Of course such unwetted fibres should be relatively easily wetted out in subsequent melt processing, since matrix and reinforcement are physically in close proximity, but this stage may require a prolonged process. It is possible that some solvent may become concentrated at the surface of the fibre, even trapped in the uneven texture and slight porosity of some fibres. When the interface is placed under tension, such trapped solvent may be released, preferentially plasticizing that critical region.

Figure 3.4 Fracture surface of solution impregnated material, showing detachment of the matrix at the fibre surface

McCullough and Pangelinan[91] argue that there is a natural tendency for molecular phase separation at a surface. The lower molecular weight elements are preferentially drawn to fill up the awkward spaces at the interface: such an interface would be less tough than would be expected of the whole polymer. It is arguable that, in a polymer solution, the molecules would be more free to phase separate in this way than in a melt. There is a growing body of evidence suggesting that, with identical polymers and fibres, solution impregnation gives less satisfactory interface control than with melts. It is at least possible that the difference between these systems does not reflect a negative aspect of solution processing but rather a positive gain from the use of melts.

Whatever the difficulties of solution prepregging, it is clear that the method has a place in a total thermoplastics picture. Because of the availablility of standard production processing technology, it offers a low cost route to the impregnation of wide fabric materials, which are often preferred for rapid stamping into deep drawn components. Although solution technology of amorphous polymers has been available for nearly two decades, it is only recently that it has become established. Solution impregnation of broadgoods provides a complementary product to melt impregnated tape products.

3.2.2 Chain extension processes

Chain extension processes are usually a subset of melt or solution impregnation. They aim to utilize the low viscosity of the prepolymer to achieve wetting of the fibres, and subsequently to achieve toughness in the matrix by linear chain extension. Resin systems that are made into linear chain extendable prepolymer prepregs designed to cure in the moulding operation (Section 3.1.2 above) could, in principle, be precured before assembly by thermoplastic techniques. In one variant of this approach the agent catalysing chain extension is directly incorporated on to the fibre surface[92] thereby directly controlling and enhancing that critical region. Certain polymers, such as polyphenylenesulfide, have a natural tendency to chain extend in the melt. There are significant benefits to be gained from controlled chain extension, but the emphasis must be on control. The termination of such reactions, and the avoidance of undesirable branching or crosslinking, are not easy, and should properly be carried out by those whose business is chemistry.

3.2.3 Melt impregnation

Melt impregnation with fully polymerized polymers is both the most difficult and the most desirable technology. The difficulty we have already identified as that of spreading highly viscous resin over a large surface area and through a tightly packed network of fibres. Theoretical calculation[93] suggests that this should take several hours, and, despite the obvious desirability of such products, the absence, until 1981, of well wetted thermoplastic composites with high loadings of fibres empirically supports that conclusion. Successful melt impregnation is neither an easy nor an obvious technology. Indeed Gosnell[94] observes that 'the difficulty of melt impregnation has been demonstrated by several unsuccessful attempts to build prepreg machines for this purpose'. Despite this, it has the advantage that, if it can be achieved, it is inherently the most direct and so the most controllable system, incorporating into the composite only those ingredients that are intended to be integral components of the finished product. McMahon[95] observed that, with the introduction of melt-impregnated carbon fibre reinforced PEEK, which product he observed to have an unprecedently high level of fibre wetting and excellent uniformity of fibre distribution, something changed in the field of thermoplastic composites. We must explore how this change was brought about.

Confronted with high viscosity resin to be impregnated, a natural reaction is to push very hard. Typically film or powder coated fibres are fed to a calender roll to force the impregnation. Such an approach suffers the disadvantage that applying pressure from both sides actually forces the fibres together, so making impregnation more difficult. A preferred alternative is to pump the resin through by applying pressure on one side of the fibrous roving, for example by urging a band of continuous fibres against a spreader surface, where the polymer is trapped in a nip between the fibres and that surface[96]. This allows the resin to flow through the fibres and open them up as it does so. Preferably this process is repeated a number of times, but with the sequential spreader surfaces on alternate sides of the roving, so that resin which had passed through the fibre is subsequently pushed back[96,97]. At these subsequent 'nips' the resin preferentially flows through the unwetted regions of the band of fibres, since this is now the path of least resistance. A variant of this nip process is to supply the resin direct to the nips though the spreader surface[98], which is designed to be porous with an internal feed of polymer. Further impregnation is sometimes achieved by drawing the impregnated tow though a tapered slot die, where high pressures may be generated to help to redistribute the resin locally. The die can also control the resin content by scraping off excess resin. High work input is almost axiomatic when working with high viscosity systems but, in the optimum process[99], any friction surfaces are always lubricated by the polymer melt, so reducing any tendency to attrition in the fibres.

A variant of this approach uses polymer of low melt viscosity[100]. Polymer melts have non-Newtonian viscosity characteristics. In order to achieve good fibre wetting, a strong preference is observed for resins with a low viscosity at low shear stress[101,102], since fibre wetting is inherently a low flow process[103]. Usually such resins have low molecular weight, and so would be expected to be rejected as not being adequately tough for service[104]. Surprisingly, however, when used as resins for composite materials, they can give excellent properties[105]. These results can be post rationalized by recourse to the arguments of fracture mechanics[106], which show that, when normally brittle materials such as glass are spun into fine fibres, thin sections are surprisingly strong. In the case of resin matrices the scale at which they are distributed is of the same scale as the fibres – a few microns only – and at such a scale even low molecular weight resins can be tough. It has been suggested[107] that the critical requirement for toughness in a composite is that the fracture mechanics parameter – the plastic zone size of the resin – is greater than the diameter of the fibre, which is taken to be equivalent to the largest resin rich region in a well-prepared fibre reinforced composite of high fibre loading. This opinion is substantiated by comparing the fracture toughness of composite materials with that of the resins from which they are known to be made[108]. Above a certain threshold of resin toughness, there is little gain in composite performance. Despite the success that can be attributed to the use of low molecular weight polymers as composite matrices, the viscosity of such polymer melts are usually several orders of magnitude greater than those of conventional thermosetting resins or polymer solutions. It is also necessary to devise some preferred method of working the resin into the fibres.

Even if successful composites can be achieved from lower molecular weight polymers, it may still be desirable to achieve direct melt impregnation with inherently tougher, high molecular weight, polymers. One way to achieve this is to incorporate a plasticizer[109,110], which can reduce the viscosity of the polymer melt, facilitating impregnation. Preferably such a plasticizer should then be removed from the material, leaving only the undiluted high molecular weight polymer as the matrix phase. Given the very great difficulty of eliminating highly volatile solvents from solution impregnated composite materials[111], the notion of extracting high boiling point plasticizers from within the constraints of an impregnated fibre bundle is implausible. Surprisingly, however, it can be achieved by using certain categories of plasticizers with melts of polyethersulphones and polyetherketones. Further, there is evidence that the preferred process for eliminating the plasticizer actually provides an unexpected method for enhancing the impregnation of the fibres and reduces the requirement for high work input. This mechanism is found to be especially advantageous when impregnating woven fibre product forms[112], where work input might otherwise cause distortion of the weft fibres in the fabric.

High viscosity lies at the heart of the problem of thermoplastic prepregging, but polymer melts undergo various degrees of shear thinning that can reduce viscosity by several orders of magnitude[113]. Processes to exploit this characteristic call for high capital investment. Gosnell[114] speculates that an intensification process comprising high shear supplemented by laser heating may be used in the impregnation of carbon fibres with PEEK.

At a more conventional level Chung and McMahon[115,116] claim to overcome the viscosity problem by the use of a pressurized crosshead extruder system. Chung emphasizes that close control must be achieved of spreading the carbon fibres before they enter the crosshead. Rapid cooling of the impregnated tape as it emerges from the die is desirable in order to suppress voids. The process is claimed to give 'fair wet out'[117]. Other workers[118] report no significant fibre wetting from such crosshead coating processes.

3.2.4 The integration of impregnation technologies

We have identified a range of approaches to the goal of producing a high quality thermoplastic material in which the fibres are well wetted by, and uniformly dispersed in, the resin. The art of optimizing technology calls for the integration of preferred elements from a diversity of options. The verb to impregnate has a range of meanings – most literally it means to make potentially fruitful. The fruits of composite manufacture are the properties derived from the structure, and by those fruits we must judge the various technologies.

3.3 A comparison of product forms

A wide range of product forms have been described for thermoplastic composites. The materials can be characterized in a number of ways (Table 3.1). We can use these characteristics to describe the range of product forms that have been discussed with respect to carbon fibre reinforced polyketones (Table 3.2).

Table 3.1 Characteristics of product forms

Feature	A		B
Feel	Boardy	or	Drapable
Anisotropy	Uniaxial	or	Woven
Organization	Collimated	or	Twisted
Fibre length	Continuous	or	Long discontinuous
Interface	Preimpregnated	or	Assembled
Status	Commercial	or	Laboratory

Table 3.2 identifies thirteen distinct product forms whose status ranges from fully qualified commercial product to laboratory curiosity (at this time). The list is not intended to be exhaustive; ingenious materials scientists will inevitably create further options. The variety of product forms provides the designer with a wide

Table 3.2 Product forms for carbon fibre polyketone

Product morphology	' Feel	Anisotropy	Org.	Length	Interface	Status
Preimpregnated tape[119]	A	A	A	A	A	A
Crimped tape[120]	A	A	B	A	A	B
Impregnated woven fabric[121]	A	B	A	A	A	B
Woven tape[122]	A/B	B	A	A	A	A
Commingled[123]	B	B (or A)	A	A	B	A
Cowoven[124]	B	B (or A)	A	A	B	A
Spun staple[125]	B	B (or A)	B	B	B	A
Powder coated fabric[126]	B	B (or A)	A	A	B	B
Powder plus tube[127]	B	B (or A)	A	A	B	B
Powder coated fibre[128]	B	A (or B)	A	A	A/B	B
Individually coated fibre[129]	B	A (or B)	A	A	A	B
Long discontinuous fibre[130]	A	A	A/B	B	B/A	A
Film stacked[131]	A	B (or A)	A	A	B	A

A/B indicates partially and (or A) implies that this is possible but not usual.

choice from which to select the optimum material for his needs. Variety is not always a boon; there is some evidence that the proliferation of product forms has also caused some confusion. It is necessary to consider some comparison to establish the criteria against which the selection process should be made.

A comparison of composite materials product forms can be made on three counts: service performance, ease of processing and cost.

With respect to cost, it is difficult, at this stage, to make any direct comparison between different product forms of the same components. It would be reasonable to suppose that the cost of preimpregnated materials should be greater than those of materials systems where the ingredients are assembled together for impregnation after shaping. This is not necessarily the case. The manufacture of fibres or fine powders that may be required to make such products are additional costly stages. When the cost of the extended processing cycle to impregnate the assembled ingredients is considered, the total cost of those products which must be impregnated after shaping may be significantly higher than their preimpregnated brothers. What is clear at the present time is that preimpregnated thermoplastic composite materials are more expensive than their thermosetting cousins of similar form. The reason for this difference is the relative difficulty of impregnating the fibres with highly viscous thermoplastic polymers compared to the ease of wetting the fibres with monomers or prepolymers of thermosetting systems. Added to this is the feature that thermoplastic structural composites are still only made in relatively low tonnages so that full economies of scale have yet to be achieved. In the final analysis it is the lifetime cost of a material that counts, and in this lifetime must be counted storage, processing, assembly, reclaimability, maintenance and repair. It is expected that lifetime costs of thermoplastics, and in particular the preimpregnated product forms, will be economically advantageous.

A major element of the final cost of a composite component is the cost of processing. One school of thought prefers drapable product forms, since these can be conveniently handled in existing hand lay up and autoclave processing technologies. One extreme of this approach is to introduce tack as well as drapability to the composite preform. This can be done, in the case of solution impregnated amorphous resins, by leaving some solvent remaining in the prepreg sufficiently to soften a fully wetted material, thereby making it both drapable and able to adhere to itself. The residual solvent can subsequently be eliminated by vacuum during the processing operation. An alternative approach, with powder impregnated fibres, is to use a sacrificial binder, first to stick the powder to the fibre, and then, when wetted, to stick the layers of powder coated materials together. The water, and preferably the binder, are subsequently eliminated during processing. A particular concern with all drapable product forms is that they are inherently bulky. This bulk must be reduced during moulding, so that removal of air becomes the key stage of processing such materials. This presents little problem for simple flat laminates, but debulking can be a significant embarrassment in consolidation of shaped structures, leading to voids or crimped fibres at corners. The issue of debulking may be of especial concern with the tube coated powder impregnated product form, where difficulty has been found in making good quality mouldings[132]. In this case much of the air to be extracted is trapped inside the tube. Hand lay up and autoclave technology are suited to the processing of large area double curvature structures when only a low production rate is required. Drapable product forms can also provide the ability to produce highly complex shapes, for example by employing textile technologies. Complex shapeability can also be enhanced by discontinuous fibres and by incorporating twist or crimp into the fibres, so that the fibre array can be stretched during forming[133]. In comparing long discontinuous fibres with crimped fibres it should be noted that, when stretched, crimped fibres straighten and so become less compliant. This leads to stable drawing. Discontinuous fibre assemblies necessarily become weaker as they are stretched, potentially giving local rupture or uneven drawing. Preimpregnated product forms are also desirable for very large mouldings, since they can be consolidated with minimum pressure: high moulding pressures may be required to wet out the fibres in the various assembled product forms and high pressures over large areas are expensive in tooling costs. It is in rapid forming, or mass production processes, that fully impregnated product forms find their natural expression. For rapid forming of complex shapes, impregnated woven fabrics, or woven tape products, are sometimes preferred. In such materials the weave form can lock in the fibre array, preventing the gross variations of ply thickness that can occur as a uniaxial tape seeks to conform to shape. Woven preimpregnated tape products can be made available in carpet widths, and have partial drapability. The fact that they need only be melted and fused in order to make a structure is the advantage of thermoplastic materials that is only found in preimpregnated product forms. Less obvious, but none the less critical, is the quality assurance in processing gained from working with preimpregnated product forms. Shape is easily perceived, but the interface between the matrix and the fibres, especially in a complex geometry moulding, is not obvious to external inspection: confidence that that critical, but invisible, component is in place as a result of preimpregnation is a desirable factor.

Service performance in composite materials primarily means mechanical performance. Comparison of mechanical performance usually means comparison of stiffness and strength in simple coupon tests. In such simple tests it is possible, but sometimes misleading, to compare data sheet values: if this is done then the test methods used must be carefully scrutinized, since some surprisingly good results are occasionally accompanied by the caveat 'modified test method'. It is therefore preferable to make direct comparisons, ideally within a single laboratory. Even here care must be taken that the specimens are moulded according to suitable procedures. Obviously comparing preimpregnated and assembled product forms moulded under a short processing cycle optimized for preimpregnated products would place simply assembled materials at a significant disadvantage. Surprisingly few direct comparisons of material product forms have been published. Measuria[134] compared three different preimpregnated products with three assembled materials. The preimpregnated products were moulded rapidly and the assembled products were given extended processing cycles to allow optimum wet out of the fibres. Measuria found that all product forms gave similar stiffness results, confirming a high level of wetting of the fibres, even in the assembled products, but that, in respect of strength measurements (Table 3.3), the assembled product forms gave significantly reduced values in comparison with their preimpregnated cousins.

Table 3.3 Strengths of CF/PEEK product forms[135, 136]

	Axial flexural strength MN/m^2	Short beam shear strength MN/m^2	Falling weight impact energy to break 2 mm sheet
Preimpregnated products			
Cross plied uniaxial	907	76	23
Woven single tow	929	68	23
Impregnated woven fabric	1,052	80	29
Impregnation after shaping			
Cowoven fibres	782	60	13
Film stacked	680	67	9
Powder coated fabric	545	54	

Inspection of the microstructure of such samples showed a good uniformity of fibre matrix distribution in the preimpregnated forms but severe non-uniformity for the assembled products.

The fibre/resin distribution, while important, is not necessarily the main cause of the degradation in strength. I believe that the high work input, in terms of high moulding pressure for a long time, necessary to wet out the fibres in the assembled product forms has caused damage to the fibres: in order to force the resin into the fibrous bed, it is necessary to force the fibres against one another. This problem can be particularly severe in processing assembled woven product forms; for preimpregnated materials each fibre is protected by a viscous layer of resin so that they are not subject to damage. Although carried out in one laboratory, the results may be considered inconclusive, since their authors had a particular interest in the preimpregnated product forms. No such objection can be levelled against the results of Silverman[137, 138] and Nardin[139], who have made the comparison of preimpregnated tape and commingled product forms, using both the standard high strength fibres and ultra high modulus pitch fibres. The results of Silverman and his colleagues (Tables 3.4 to 3.6) confirm that the various assembled product forms have approximately 70% of the strength of preimpregnated products.

Table 3.4 Comparison of product forms of CF/PEEK based on high-strength (AS4) fibres[140]

| Product form | Axial flexure | | Transverse tension |
	Modulus GN/m²	Strength MN/m²	Strength MN/m²
Preimpregnated tape	108	1,687	91
Commingled fibres			
no-crimp structure	106	1,222	26
standard	98	1,514	64
Cowoven fibres	65	1,150	Not tested

Table 3.5 Comparison of product forms of CF/PEEK based on high-strength (AS4) fibres[141]

| Product form | Axial tension | | Axial compression | | Transverse tension | |
	Modulus GN/m²	Strength MN/m²	Modulus GN/m²	Strength MN/m²	Modulus GN/m²	Strength MN/m²
Preimpregnated tape	132	>1,172	113	1,175	8.9	91
Commingled fibres	127	> 782	115	865	8.9	64

Table 3.6 Comparison of product forms of CF/PEEK based on ultra high modulus (P75) fibre[142]

| Product form | Longitudinal tension | | Transverse tension | |
	Modulus GN/m²	Strength MN/m²	Modulus GN/m²	Strength MN/m²
Preimpregnated tape	303	808	5.9	21
Commingled fibres	230	>600	7.8	11

The observation of low stiffness in the assembled products based on ultra high modulus fibres emphasizes the difficulty of working with such materials. Even though Silverman followed the manufacturer's recommended processing conditions, there may be some inadequacy of fibre wetting in those mouldings.

Silverman also shows comparative micrographs of the different product forms, confirming the relative non-uniformity of resin distribution in the mouldings of assembled product forms.

Nardin[143] also observes approximately 70% strengths in longitudinal and transverse tensile tests on uniaxial commingled products, compared to preimpregnated materials. The former he attributes to difficulty in controlling fibre orientation and the latter to inferior interface properties in the assembled products.

In comparisons between different drapable products, proponents of the spun staple yarn form claim that the twist in the product locks in the resin fibre distribution, giving a more uniform result than simple commingled products[144,145]. Because of the twist in the spun staple yarn product, it cannot produce the highest stiffness and strength in simple coupon testing, and in such laboratory trials it appears inferior except in respect to interlaminar shear strength, which is enhanced by the twist and the 'hairiness' of the product[146]. Morales[147], an independent observer, has demonstrated that in complex braided structures in flexure the spun staple yarn form gives at least equivalent and possibly superior properties to the commingled product despite having lower coupon properties (Table 3.7).

Table 3.7 Comparison of spun staple yarns and commingled fibre product forms[148]

	Tension				Flexure			
	Uniaxial		3-D braided		Uniaxial		3-D braided	
	Modulus GN/m^2	Strength MN/m^2	Modulus GN/m^2	Strength MN/m^2	Modulus GN/m^2	Strength MN/m^2	Modulus GN/m^2	Strength MN/m^2
Commingled	134	1,531	87	586	100	841	57	345
Spun Staple	76	834	66	441	90	724	55	414

From the experience of thermosetting composite materials, it is usually assumed that woven product forms provide lower stiffness and strength than laminated tape products. Such deterioration is presumed to be especially significant in compression strength, because of the inherent crimp in the fibres. In the case of thermoplastics Measuria[149] demonstrates that this need not necessarily be the case: all preimpregnated product forms give essentially similar performance. This result is especially confirmed for the woven preimpregnated tape product forms[150]. These arguments suggest that all product forms that tend to give enhanced drapability suffer some degradation of properties in comparison to the boardy preimpregnated forms. Well preimpregnated products all give the same high utilization of the inherent fibre properties.

One strong advantage foreseen for filamentaceous product forms is the ability to weave or braid 3-D reinforced materials. This is expected to give increased delamination resistance and enhanced toughness[151,152]. Taske and Majidi[153] have made a comparative study of laminated preimpregnated tapes, two dimensionally woven laminates and three dimensionally braided structures. They observe that all the material forms absorb a similar amount of energy on impact, and that the woven and braided forms delaminate less. In respect of residual properties after impact, as measured by flexural strength, the response of all product forms is very similar near the point of impact. Away from that point of impact the simple laminate based on preimpregnated tape give markedly superior performance (Figure 3.5).

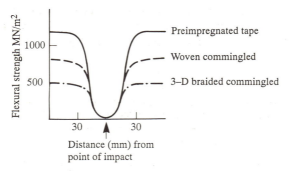

Figure 3.5 Comparison of residual strength after impact for various product forms[154]

Brandt[155] and Ko[156] have measured compression strength after impact on braided commingled products having a nominal fibre orientation of about ±20°, and have obtained similar values to those for simple quasi-isotropic laminates. The transverse compression strength after impact on those anisotropic braided structures has not been reported, but might reasonably be expected to be significantly lower than for quasi-isotropic panels, since the fibre orientation would effectively be ±70°. At this time there is no evidence that, with tough matrix resins, there is sufficient gain from 3-D reinforcement to offset the reduction in stiffness and strength.

It is insufficient to choose preimpregnated products simply if optimum mechanical performance is the goal. Which preimpregnation route is selected is also important. In particular there is sometimes the option to choose between a melt and a solution impregnation process. In the case of semi-crystalline polymers the luxury of this option does not always exist, because such polymers cannot readily be rendered into solution. This is particularly noted to be the case with polyetheretherketone. By contrast, amorphous thermoplastics can usually be dissolved readily, so that matrices such as polyethersulphone and polyetherimide can be conveniently prepared as low viscosity solutions containing about 10% by weight of polymer[157]. Such resins can also be prepregged by the melt techniques appropriate for their less tractable semi-crystalline brothers. The first advantage for solution impregnation is that there exists conventional equipment, designed to work with thermosetting resins, from which to make the prepregs. In such processes the fibres are easily wetted by surface tension forces: there is no need for specialized high temperature, high work input, technology to cope with high viscosity thermoplastic melts. Solution processing should therefore provide the less expensive option. On the negative side we have noted[158] that elimination of the last residues of solvent can be particularly difficult, and that residual solvent can lead to significant reduction in the glass transition temperature of the composite. This issue has been extensively studied[159], and several amorphous polymer prepregs, which are presumed to have been made by a solution process, are currently described as solvent free. It is therefore instructive to compare the service performance of optimum solution and melt impregnated materials. Wu[160] compares the properties of two amorphous resin composites with identical matrices and reinforcements. The axial, fibre dominated, properties are virtually identical, but the melt impregnated sample gives significantly superior properties in the transverse, or matrix dominated, response (Table 3.8).

Table 3.8 Comparative properties of solution and melt preimpregnated thermoplastic composites based on polyethersulphone of low and high molecular weight[161]

Resin	Flexural properties							
	Solution impregnation				Melt impregnation			
	Longitudinal		Transverse		Longitudinal		Transverse	
	Modulus GN/m²	Strength MN/m²	Modulus GN/m²	Strength MN/m²	Modulus GN/m²	Strength MN/m²	Modulus GN/m²	Strength MN/m²
Low Mol. Wt.	137	1,335	8	48	130	1,280	8	94
High Mol. Wt.	132	1,431	9	55	128	1,390	8	95

This result is further substantiated by Drzal[162]. He observed that, in single fibre pull out tests, samples prepared by melt coating fibres give superior shear strength to solution coated materials. These results are particularly surprising, because it would be natural to assume that more perfect wetting should be achieved by solution preimpregnation. An explanation for these observations has yet to be agreed, but several hypotheses have been advanced: detachment of the interface during drying, molecular phase separation giving a low molecular weight interphase, and trapped solvent at the interface. These issues are considered in Section 3.2.1 and Appendix 7. A more positive explanation could be that the interface in solution preimpregnated materials is not weaker than it should be, but that the interface in optimally melt impregnated materials is surprisingly good. With a number of university groups currently studying this area we may hope to receive an answer to this conundrum. In the meantime experience indicates that, for whatever reason, melt preimpregnated product forms are superior.

3.4 A selection of product forms in carbon fibre/PEEK

The wide range of choice of product form can be viewed as a boon, but excessive choice can also impede decision making. Any selection from the thirteen options listed in Table 3.2 is necessarily a subjective one. My personal selection of three

Table 3.9 Recommended product forms

Product form	Application indications
Preimpregnated tape (melt impregnation) Supplier: ICI Fiberite[163]	Optimum service performance. Well consolidated material. Maximum design freedom from use of uniaxial product. Rapid fabrication at low pressure. Demonstrated ability to form double curvature structures. Maximum quality assurance.
Woven preimpregnated tape Supplier: Quadrax Corporation[164]	Availability of broad goods. Modest drapability. Woven form to constrain excessive flow in deep drawn mouldings. Product quality determined by preimpregnated tape.
Spun staple yarn Supplier: Courtaulds[165]	Maximum compliance for construction as textile preforms. Twist allows some control of fibre–resin distribution. Potential for thin gauge product.

product forms that should meet all needs is shown in Table 3.9. I make that recommendation on the basis of technical performance and availability.

Accepting the desirability of a drapable product form to take advantage of textile technology preforms, I prefer the spun staple yarn. The fact that this gives lower coupon properties than the commingled product is offset by the better translation of those properties into complex mouldings[166]. An additional bonus for this product form is that, at the spinning stage, it can be drawn into a low denier tow, which can subsequently be woven to provide thin laminates that, by other routes, could be prohibitively expensive.

The broadgoods with limited drapability available from the woven preimpregnated single tow tape combine an excellent property profile with the ability to lay down large areas of material, and therefore are well suited to large structures of modest curvature. The constraint provided by the weave structure can usefully be exploited in the manufacture of deep drawn structures. Undoubtedly it would be desirable to have impregnated woven fabrics, with tight weave of small tows, as a preferred form for deep drawing of small complex structures, but these are not at present (1990) commercially available as PEEK matrix composites, although they are available for amorphous resin composites based on polyethersulphone and polyetherimide.

Melt preimpregnated uniaxial tape products provide the basis for optimum laminate design with the best possible properties. This product form has been extensively characterized and demonstrated in a range of mouldings of varying complexity of geometry. In respect of the interface and microstructure it is the material system that has been most specifically designed to fit its purpose, and therefore gives the greatest quality assurance. Indeed, in respect of the open literature, we probably know more about the performance of this carbon fibre reinforced polyetheretherketone and its dependence on microstructure than we do about any other available forms of composite material, including those which have been in service for twenty years. It is the detailed microstructure of this product form that we must examine in the next chapter.

References

3–1   F. N. COGSWELL, 'Rheology and Structure', in *Polymer Melt Rheology: A Guide for Industrial Practice,* Chapter 4, George Godwin Ltd (1981).

3–2   K. J. AHN, J. C. SEFERIS AND L. LETTERMAN, 'Autoclave Infusion Process: Analysis and Prediction of Resin Content', *SAMPE Quarterly,* **21**, 2, pp. 3–10 (1990).

3–3   M. L. ORTON, 'A New Resin System for Increased Productivity in Fibre Composite Processing', Internal Conference on Fibre reinforced Composites, Liverpool (1984).

3–4   I. MARCHBANK, 'Automated RTM for an Airframe Component', in *Materials and Processing – Move into the 90's,* edited by S. Benson, T. Cook, E. Trewin and R. M. Turner, Elsevier (1989).

3–5   G. LUBIN AND S. J. DASTIN, 'Aerospace Application of Composites', in *Handbook of Composites,* edited by G. Lubin, p. 740, Van Nostrand Reinhold (1982).

3–6   W. HILAKOS AND D. J. PATTERSON, 'Poltrusion Apparatus and Method for Impregnating Continuous Lengths of Multifilament and Multifibre Structures', EP 032 654 A2 (1989).

3–7   H. M. COLQUHOUN, C. C. DUDMAN, M. THOMAS, C. A. MAHONEY AND D. J. WILLIAMS, 'Synthesis, Structure and Ring-Opening Polymerization of Strained Macrocyclic Biaryls: a New Route to High Performance Materials', (in press 1990).

3–8   G. WIEDEMANN AND H. ROTHE, 'Review of Prepreg Technology', in *Developments in Reinforced Plastics,* 5th edition, G. Pritchard, pp. 83–120, Elsevier (1980).

3–9   J. A. PEACOCK AND F. N. COGSWELL, 'Fibre Reinforced Advanced Structural Composites', in I. Miles and S. Rostami, *Multiphase Polymer Systems,* Longman (in press 1990).

3–10 G. H. HARDESTY, 'Poly (Amide-Imide)/Graphite Advanced Composites', Aerospace Congress and Exposition, Longbeach (1984).

3–11 H. H. GIBBS, 'K-Polymer; a New Experimental Thermoplastic Matrix Resin for Advanced Structural Aerospace Composites', 29th National SAMPE Symposium, Reno (1984).

3–12 H. H. GIBBS, 'Processing Studies on K-Polymer Composite Materials', 30th National SAMPE Symposium, pp. 1585–1601 (1985).

3–13 P. A. HOGAN, 'The Production and Uses of Film Stacked Composites for the Aerospace Industry', SAMPE Conference (1980).

3–14 D. J. LIND AND V. J. COFFEY, British Patent 1,485,586 (1977).

3–15 L. N. PHILLIPS AND D. J. MURPHY, British Patent 1,570,000 (1980).

3–16 W. I. LEE AND G. S. SPRINGER, 'A Model of the Manufacturing Process of Thermoplastic Matrix Composites', *J. Comp. Mats,* **21**, 11, pp. 1017–1055 (1987).

3–17 A. A. STORI AND E. MAGNUS, 'An Evaluation of the Impact Properties of Carbon Fibre Reinforced Composites with Various Matrix Materials', from *Composite Structures 2,* edited by I. H. Marshall, Paisley College of Technology, pp. 332–348 (1983).

3–18 J. T. HARTNESS, 'Polyether-etherketone Matrix Composites', 14th National SAMPE Technical Conference, **14**, pp. 26–43 (1982).

3–19 R. K. OKINE, 'Analysis of Forming Parts from Advanced Thermoplastic Composite Sheet Materials', *SAMPE Journal,* **25**, 3, pp. 9–19 (1989).

3–20 A. P. PERRELLA, 'Designing Aeropsace Structures with Du Ponts LDF Thermoplastic Composites', 21st International SAMPE Technical Conference, pp. 705–719 (1989).

3–21 A. J. BARNES AND J. B. CATTANACH, 'Advances in Thermoplastic Composite Fabrication Technology', Materials Engineering Conference, Leeds (1985).

3–22 J. F. PRATTE, W. H. KRUEGAR AND I. Y. CHANG, 'High Performance Thermoplastic Composites with Poly (Ether Ketone Ketone) Matrix', 34th International SAMPE Symposium, pp. 2229–2242 (1989).

3–23 C. J. A. SLATER, 'Woven Fabrics of Carbon or Glass Fibres with Thermoplastic Polymer suitable for Fabricating in Reinforced Composites', Research Disclosure 20239 (1981).

3–24 M. E. KETTERER, 'Thermoplastic/Carbon Fiber Hybrid Yarn', NASA Contractor Report 3849, contract NASI-15749 (1984).

3–25 M. HEYM, E. D. WESTERN AND M. WINKLER, 'New Material Forms for Manufacture of Thermoplastic Composites', *Proc. PRI Conf. 'Automated Composites '88'',* Leeuwenhorst, 3/1–3/12 (1988).

3–26 A. C. HANDERMANN, 'Advances in Commingled Yarn Technology', 26th Int SAMPE Technical Conference, pp. 681–688 (1988).

3–27 L. E. TASKE AND A. P. MAJIDI, 'Performance Characteristics of Woven Carbon/PEEK Composites', *Proc. Am. Soc. for Composites,* 2nd Tech Conf., Technomic Publishing Co. (1987).

3–28 A. P. MAJIDI, M. J. ROTERMUND AND L. E. TASKE, 'Thermoplastic Preform Fabrication and Processing', *SAMPE J.,* **24**, 1, pp. 12–17 (1988).

3–29 T. LYNCH, 'Thermoplastic/Graphite Fiber Hybrid Fabrics', *SAMPE J.,* **25**, 1, pp. 17–22 (1989).

3–30 M. FUJII AND H. INOGUCHI, 'Good Wettability of Co-Woven Unidirectional Fabrics', 21st International SAMPE Technical Conference, pp. 496–505 (1989).

3–31 J. VOGELSANG, G. GREENING, R. NEUBERG AND A. HANDERMANN, 'Hybrid Yarns of HT-Thermoplastics and Carbon Fibers: a New Material Form for Advanced Composites', Chemiefassern/Textilindustrie, CT1 **39**/91 pp. T224–T228 (1989).

3–32 F. KO, P. FANG AND H. CHIU, '3-D Braided Commingled Carbon Fibre/PEEK Composites', 33rd International SAMPE Symposium, pp. 899–911 (1988).

3–33 T. J. WHITNEY AND T. W. CHOU, 'Modelling of Elastic Properties of 3-D Textile Structural Composites', *Proc. Am. Soc. for Comp., 3rd Technical Conf.,* pp. 427–436, Technomic Publishing Co. (1988).

3–34 F. KO, H. CHU AND E. YING, 'Damage Tolerance of 3-D Braided Intermingled Carbon/PEEK Composites', *Proc. 2nd Conf. on Advanced Composites,* Dearborn (1986).

3–35 J. BRANDT, T. PRELLER AND K. DRECHSLER, 'Manufacturing and Mechanical Properties of 3-D Fibre Reinforced Composites', in *Materials and Processing – Move into the 90's, Proc. 10th Int. Eur. Chapter Conf. of SAMPE,* Materials Science Monographs, **55**, pp. 63–74, Elsevier (1989).

3–36 C. T. HUA AND F. K. KO, 'Properties of 3D Braided Commingled PEEK/Carbon Composites', 21st International SAMPE Technical Conference, pp. 688–699 (1989).

3–37 MAJIDI *et al., op. cit.*

3–38 S. H. OLSON, 'Manufacturing with Commingled Yarns, Fabrics and Powder Prepreg Thermoplastic Composite Materials', 35th International SAMPE Symposium, pp. 1306–1319 (1990).

3–39 P. J. IVES AND D. J. WILLIAMS, 'Ultra-Fine High Performance Yarns in Carbon and Other Fibres: A Route to Lightweight Fabrics and Thermoplastic Prepregs', 33rd International SAMPE Symposium, pp. 858–869 (1988).

3–40 R. J. COLDICOTT, T. LONGDON, S. GREEN AND P. J. IVES, 'Heltra – A New System for Blending Fibres and Matrix for Thermoplastic – Based High Performance Composites', 34th International SAMPE Symposium, pp. 2206–2216 (1989).

3–41 R. J. COLDICOTT AND T. LONGDON, 'A New Family of High Performance Yarns for Composite Applications', in *Materials and Processing – Move into the 90's, Proc. 10th Int. Eur. Chapter Conf. of SAMPE,* Materials Science Monographs, **55**, pp. 359–370 Elsevier, (1989).

3–42 R. M. JOURNAL, 'Blended Spun Yarns for Materials of High Mechanical and Thermal Resistance', PRI Conference, Liverpool, 21 February 1990.

3–43 See 3–41.

3–44 See 3–42.

3–45 See 3–41.

3–46 J. D. RUSSELL, 'Comparison of Processing Techniques for Filmix Unidirectional Commingled Fabric', 35th International SAMPE Symposium, pp. 13–24 (1990).

3–47 A. MORALES, J. N. CHU, P. FANG AND F. K. KO, 'Structure and Properties of Unidirectional and 3-D Braided Commingled PEEK/Graphite Composites SPE, 47th ANTEC, pp. 1459–1564 (1989).

3–48 See 3–34.

3–49 See 3–35.

3–50 L. E. TASKE AND A. P. MAJIDI, 'Parametric Studies of Impact Testing of Laminated and Woven Archetextures', *Proc. Am. Soc for Composites 3rd Technical Conf.,* pp. 374–383, Technomic Publishing Co. (1988).

3–51 R. V. PRICE, 'Production of Impregnated Roving', US Patent 3,742,106 (1973).

3–52 K. KENDALL, 'The Impossibility of Comminating Small Particles of Compression', *Nature,* **272**, pp. 5655, 710–711 (1978).

3–53 D. J. BLUNDELL, P. J. MEAKIN AND F. N. COGSWELL, 'Process for Making Microporous Products and the Products thereof', EPO 0297 744 (1989).

3–54   J. MUZZY, B. VARUGHESE, V. THAMMONGKOL AND W. TINCHER, 'Electrostatic Prepregging of Thermoplastic Matrices', *SAMPE Journal*, **25**, 5, pp. 15–21 (1989).

3–55   See 3–34.

3–56   B. VARUGHESE, J. MUZZY AND R. M. BAUCOM, 'Combining LaRC-TPI Powder with Carbon Fiber by Electrostatic Fluidized Bed Coating', 21st International SAMPE Technical Conference, pp. 536–543 (1989).

3–57   J. MUZZY, B. VARUGHESE AND P. H. YANG, 'Flexible Towpreg by Powder Fusion Coating of Filaments', SPE ANTEC (1990).

3–58   See 3–55.

3–59   J. L. THRONE AND M.-S. SOH,, 'Electrostatic Dry Powder Prepregging of Carbon Fibre', 35th International SAMPE Symposium, pp. 2086–2101 (1990).

3–60   R. GANGA, 'Fibre Impregnee Thermoplastique (FIT)', *Composites et Noveaux Materiaux*, 5 May 1984.

3–61   M. THIEDE-SMET, 'Study of Processing Parameters of PEEK/Graphite Composite Fabricated with 'FIT' Prepreg', 34th International SAMPE Symposium, pp. 1223–1234 (1989).

3–62   J. T. HARTNESS, 'Thermoplastic Powder Technology for Advanced Composite Systems', 33rd International SAMPE Symposium, pp. 1458–1472 (1988).

3–63   J. T. HARTNESS, 'Thermoplastic Powder Technology for Advanced Composite Systems', *J. Thermoplastic Composite Materials*, **1**, pp. 210–220 (1988).

3–64   S. CLEMANS AND J. T. HARTNESS, 'Thermoplastic Prepreg Technology for High Performance Composites', *SAMPE Quarterly*, **20**, 4, pp. 38–42 (1989).

3–65   I. Y. CHANG, 'Thermoplastic Matrix Continuous Filament Composites of Kevlar Aramid or Graphite Fiber', *Composites Science and Technology*, **24**, pp. 61–79 (1985).

3–66   F. N. COGSWELL, 'The Processing Science of Thermoplastic Structural Composites', *International Polymer Processing*, **1**, 4, pp. 157–165 (1987).

3–67   D. H. BOWEN, N. J. MATTINGLEY AND R. J. SAMBUL, UK Patent 'Improvements in or Relating to Apparatus for Coating Fibres', 1,279,252 (1968).

3–68   R. AND J. DYKESTERHOUSE, 'Production of Improved Preimpregnated Material Comprising a Particulate Thermoplastic Polymer suitable for use in the Formation of a Substantially Void-Free Fiber-Reinforced Composite Articles', WO 88/03468 (1988).

3–69   J. FISCHER, H. ZEINER, D. NISSEN, G. HEINZ, EPO 0177701 (1985).

3–70   G. DE JAGER, 'Method and Apparatus for Applying Powdered Materials to Filaments', WO 87/00563 (1987).

3–71   H. KOSUDA, Y. NAGAIZUMI-CHO AND Y. ENDOH, 'Method for Producing Carbon Fibre Reinforced Thermoplastic Resin Product', EPO 0272 648 (1988).

3–72   E. MOREL, G. RICHERT, 'Materiaux Composites a Matrice Thermo-Plastique et Fibres Longues', *Annales des Composites*, **3**, pp. 5–30 (1983).

3–73   D. HODGE, private communication (1989).

3–74   G. R. BELBIN, I. BREWSTER, F. N. COGSWELL, D. J. HEZZELL AND M. S. SWERDLOW, 'Carbon Fibre Reinforced PEEK: A Thermoplastic Composite for Aerospace Applications', STRESA Meeting of SAMPE (1982).

3–75   N. TURTON AND J. McAINSH, 'Thermoplastic Compositions', US Patent 3,785, 916 (1974).

3–76   J. McAINSH, 'The Reinforcement of Polysulphones and other Thermoplastics with Continuous Carbon Fibre', BPF 8th International Reinforced Plastics Congress (1972).

3–77   N. J. JOHNSTON, T. K. O'BRIEN, D. H. MORRIS AND R. A. SIMONDS, 'Interlaminar Fracture Toughness of Composites II: Refinement of the Edge Delamination Test and Application to Thermoplastics', 20th National SAMPE Symposium, pp. 502–517 (1983).

3–78   J. K. GOODMAN AND A. R. LOOS, 'Thermoplastic Prepreg Manufacture', *Proc. Am. Soc. for Composite Materials*, 4th Tech. Conf., pp. 746–754, Technomic (1989).

3–79   S. L. PEAKE, 'Processes for the Preparation of Reinforced Thermoplastic Composites', USP 4,563,232 (1986).

3–80   J. E. HARRIS AND M. J. MICHNO, 'Thermoplastic Composites Comprising a Polyaryl Ether Sulphone Matrix Resin', EPO 0303 337 (1989).

3–81   M. S. SEFTON, P. T. MCGRAIL, J. A. PEACOCK, S. P. WILKINSON, R. A. CRICK, M. DAVIES AND G. ALMEN, 'Semi-Interpenetrating Polymer Networks as a Route to Toughening of Epoxy Resin Matrix Composites', 19th Int. SAMPE Technical Conf., **19**, pp. 700–710 (1987).

3–82   See 3–64.

3–83   See 3–77.

3–84   See 3–78.

3–85   D. LEESER AND B. BANISTER, 'Amorphous Thermoplastic Matrix Composites for New Applications', 21st International SAMPE Technical Conference, pp. 507–513 (1989).

3–86   L. POST AND W. H. M. VAN DREUMEL, 'Continuous Fiber Reinforced Thermoplastics', in *Progress in Advanced Materials and Processes: Durability, Reliability and Quality Control'*, edited by G. Bartelds and R. J. Schiekelmann, pp. 201–210 Elsevier (1985).

3–87   S. PEAKE, A. MARANCI, D. MEGNA, J. POWERS AND W. TRZASKOS, 'Processing and Properties of Polyetherimide Composites', edited by S. Benson, T. Cook, E. Trewin and R. M. Turner in *Materials and Processing – Move into the 90's,* Elsevier, 111–119 (1989).

3–88   See 3–85.

3–89   G. M. WU, AND J. M. SCHULTZ, D. J. HODGE AND F. N. COGSWELL, 'Solution Impregnation of Carbon Fiber Reinforced Poly (ethersulphone) Composites', SPE Antec, pp. 1390–1392 (1990).

3–90   L. DRZAL, results presented at symposium 'Tailored Interfaces in Composite Materials', Boston (1989).

3–91   R. McCULLOUGH AND L. PANGELINAN, private communication (1989).

3–92   M. V. WARD, E. NIELD AND P. A. STANILAND, 'Method of Increasing Molecular Weight of Poly(aryl ethers)', EP 0125816B1 (1987).

3–93   See 3–16.

3–94   R. B. GOSNELL, 'Thermoplastic Resins', in *Engineered Materials Handbook Volume 1: Composites,* edited by T. J. Reinhart, pp. 97–104 (1987).

3–95   P. E. McMAHON, 'Thermoplastic Carbon Fibre Composites', printed in *Development in Reinforced Plastics – 4,* edited by G. Pritchard, pp. 1–30 (1984).

3–96   F. N. COGSWELL, D. J. HEZZELL AND P. J. WILLIAMS, 'Fibre-Reinforced Compositions and Methods for Producing such Compositions', USP 4,549, 920 (1981).

3–97   H. BIJSTERBOSCH AND R. J. GAYMANS, 'Pultrusion Process: Nylon 6/glass fibres', *Fibre Reinforced Composites FRC '90,* I. Mech. E. (1990).

3–98   E. K. BINNERSLEY, W. H. KRUEGER AND A. J. JONES, 'Method of Making a Fiber Reinforced Thermoplastic Materials and Product Obtained thereby', EPO 0167 303 (1986).

3–99   See 3–96.

3–100 *Ibid.*

3–101 *Ibid.*

3–102 K. F. WISSBRUN, R. D. ORWELL, 'Method for the formation of composite materials of thermoplastic liquid crystalline polymers and articles produced thereby', EPO 0117098 (1984).

3–103 F. N. COGSWELL, *Polymer Melt Rheology: A Guide for Industrial Practice',* George Godwin (1981).

3–104 See 3–95.

3–105 COGSWELL, HEZZELL AND WILLIAMS, *op. cit.*

3–106 F. N. COGSWELL, 'Microstructure and Properties of Thermoplastic Aromatic Polymer Composites', 28th National SAMPE Symposium, pp. 528–534 (1983).

3–107 *Ibid.*

3–108 N. J. JOHNSTON AND P. M. HERGENROTHER, 'High Performance Thermoplastic: A Review of Neat Resin and Composite Properties', 32nd International SAMPE Symposium (1987).

3–109 F. N. COGSWELL AND P. A. STANILAND, 'Method of Producing Fibre Reinforced Composition', USP 4,541,884 (1985).

3–110 F. N. COGSWELL, E. NIELD AND P. A. STANILAND, 'Method of Producing Fibre Reinforced Composition', EPO 0,102,158 (1983).

3–111 See 3–78.

3–112 F. N. COGSWELL AND U. MEASURIA, 'Reinforced Fibre Products and Process of Making', USP 4,624,886 (1988).

3–113 See 3–94.

3–114 See 3–94.

3–115 T. S. CHUNG AND P. E. McMAHON, 'Thermoplastic Polyester Amide–Carbon Fiber Composites', *J. Am. Polymer Sci.,* **31**, pp. 965–977 (1986).

3–116 T. S. CHUNG, H. FURST, Z. GURION, P. E. McMAHON, R. D. ORWELL AND D. PALANGIO, 'Process for Preparing Tapes from Thermoplastic Polymers and Carbon Fibres', USP 4,588, 538 (1986).

3–117 See 3–115.

3–118 See 3–97.

3–119 D. C. LEACH AND P. SCHMITZ, 'Product Forms in APC Thermoplastic Matrix Composites', 34th International SAMPE Symposium, pp. 1265–1274 (1989).

3–120 D. BUTLER AND F. N. COGSWELL, unpublished work.

3–121 MEASURIA AND COGSWELL, 'Aromatic Polymer Composites: Broadening the Range', *SAMPE Journal*, **21**, 5, pp. 26–31 (1985).

3–122 F. COLUCCI, 'Quadrax Interlaced Innovations', *Aerospace Composites and Materials,* **1**, 6 (1989).

3–123 See 3–31.

3–124 U. MEASURIA AND F. N. COGSWELL, 'Thermoplastic Composites in Woven Fabric Form', I.Mech.E C21/86, pp. 1–6, (1986).

3–125 See 3–39.

3–126 See 3–65.

3–127 See 3–61.

3–128 See 3–56.

3–129 D. J. HEZZELL AND F. N. COGSWELL, unpublished work.

3–130 See 3–22.

3–131 See 3–18.

3–132 See 3–61.

3–133 F. N. COGSWELL AND D. C. LEACH, 'Processing Science of Continuous Fibre Reinforced Thermoplastic Composites', *SAMPE Journal,* **24**, 4, 33–42 (1988).

3–134 See 3–124.

3–135 See 3–124.

3–136 See 3–66.

3–137 E. M. SILVERMAN AND R. J. JONES, 'Property and Processing Performance of Graphite/PEEK Prepreg Tapes and Fabrics', *SAMPE Journal,* **24**, 4, pp. 33–42 (1988).

3–138 E. SILVERMAN, R. A. CRIESE AND W. C. FORBES, 'Property Performance of Thermoplastic Composites for Spacecraft Systems', *SAMPE Journal,* **25**, 6, pp. 38–47 (1989).

3–139 M. NARDIN, E. M. ASLOUN, J. SCHULTZ, J. BRANDT AND H. RICHTER, 'Investigations on the Interface and Mechanical Properties of Composites based on Non-Impregnated Carbon Fibre Reinforced Thermoplastic Preforms', *Proc. 11th International European Chapter of SAMPE,* pp. 281–292 (1990).

3–140 See 3–137.

3–141 See 3–138.

3–142 See 3–138.

3–143 See 3–139.

3–144 See 3–40.

3–145 See 3–41.

3–146 See 3–40.

3–147 See 3–47.

3–148 See 3–47.

3–149 See 3–121.

3–150 See 3–122.

3–151 See 3–35.

3–152 See 3–50.

3–153 *Ibid.*

3–154 *Ibid.*

3–155 See 3–35.

3–156 See 3–34.

3–157 See 3–75.

3–158 See 3–77.

3–159 See 3–78.

3–160 See 3–89.

3–161 See 3–89.

3–162 See 3–90.

3–163 See 3–119.

3–164 See 3–122.

3–165 See 3–41.

3–166 See 3–47.

4 The microstructure of Aromatic Polymer Composites

We have established that the properties of composite materials derive from their components and how those components are integrated. It is the integration process that determines the organization of the components and so creates the microstructure on which the properties ultimately depend. It is in the nature of composite materials that they are heterogeneous systems, and the level of that hetrogeneity depends upon the scale at which the material is examined. Ultimately all macroscopic deformations are determined by microstructural processes, and, if we are to understand the service performance of a composite structure, we must first appreciate the microstructure of the material.

The microstructure can be considered in three parts: the organization of the reinforcing fibres and matrix; the texture, or morphology, of the matrix phase; and the residual strains within the structure. There are also the internal structure of the reinforcing fibre and the molecular architecture of the matrix polymer; those are defined by the choice of ingredients and are described in Chapter 2. This section considers only those aspects of the microstructure that are controlled by the composite manufacturing process. As a vehicle for carrying forward this study, the continuous fibre reinforced tape product preimpregnated with polyetheretherketone is both the simplest and best characterized system.

4.1 Voids

Before considering the detail of the microstructure of composite materials, it is necessary to consider first any absence of detail, or voids, in the structure. Voids may exist within the preimpregnated tape or ply, and potential voids certainly exist at the surface of the plies when they are rough. These irregularities can transpose into voids in a consolidated laminate. Interply voids, which are flat in profile, are readily detected by ultrasonic C-Scan. Large intra-ply voids can also show up in C-Scan. Small voids or cracks can sometimes be traced by using dye penetrants or water. The presence of water in such flaws can be detected by nuclear magnetic resonuance[1]. It is also possible, in principle, to use a sophisticated density measurement to estimate void content. Using this approach, Manson and Seferis[2] suggested that fast cooled mouldings were more prone to voiding. The approach requires very accurate knowledge of fibre content and crystallinity, which both affect density independently of void content. Such knowledge is not always readily available. Microscopic inspection of sectioned laminates is the ultimate test for voids. Figures 4.1 and 4.2 show typical mouldings from melt impregnated carbon fibre reinforced PEEK.

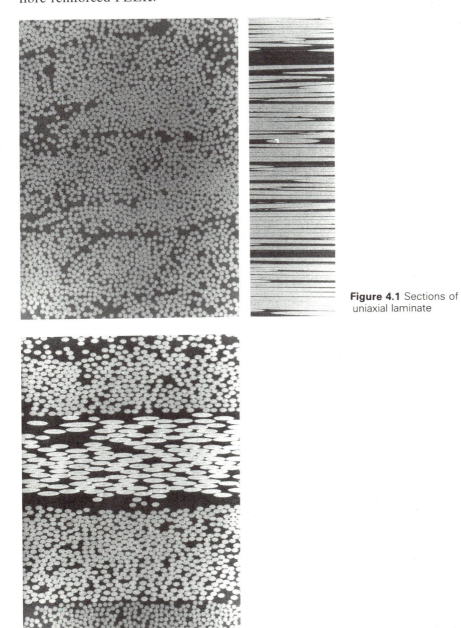

Figure 4.1 Sections of uniaxial laminate

Figure 4.2 Section of quasi-isotropic laminate

For melt impregnated thermoplastic composites, when they are correctly processed, the void content is usually considered to be less than 0.25%. Higher values are observed on mouldings made below the recommended temperature[3] and on other product forms[4]. It is seldom justified to quote absence of data in verification of a claim but, in composite materials such as APC-2, voids are virtually never referred to except to remark on their absence[5]. This helps to substantiate the view that such features are not significant in those materials.

4.2 Organization of the fibres

In this section we must consider how the fibres are distributed in the matrix and how they are aligned.

4.2.1 Fibre distribution

A typical sample of carbon fibre reinforced polyetheretherketone contains 68% by weight of high-strength carbon fibre. In the solid state this translates into a volume fraction of 61% and in the melt state 59% (Table 4.1).

Table 4.1 Volume fraction of fibre

Component	Weight fraction	Specific gravity		Volume fraction	
		@ 23°C	@ 380°C	@ 23°C	@ 380°C
Carbon fibre	0.68	1.76	1.74	0.61	0.59
PEEK resin	0.32	1.30	1.08	0.39	0.41

The small difference between ambient service temperature and the typical processing temperature is an important one as we shall see when we consider the processability of these compounds. The change in volume fraction is also significant when the internal stresses within the composite due to temperature changes are considered. The selection of 60–62% by volume as the 'standard' volume fraction for fibre reinforced composites at service temperature has been arrived at empirically. Higher fibre fractions generally give difficulty in obtaining void free processing and low translation of the inherent stiffness and strength of the fibre are achieved: there is little gain in mechanical performance to offset the processing difficulties. By contrast, low fibre fractions decrease the mechanical performance in proportion to the fibre fraction and, if excess resin is present, controlling resin fibre distribution can be a problem. Lower fibre fractions are sometimes considered desirable, for example in woven fabric systems where neat resin may be called upon to fill in the interstices of the weave. Since all prepreg systems naturally have a resin rich surface layer (see below) and that surface layer is comparable in thickness to the diameter of the fibres, thinner prepregs, or prepregs based on larger diameter fibres, are sometimes designed to a lower optimum fibre content. At this time I know of no theoretical analysis which expresses the optimum fibre content for a composite material, and must therefore accept the judgement of experience.

The reinforcing fibres are distributed non-uniformly in the composite. Ideally we might seek a uniform hexagonal packing of fibres (Figure 4.3a), so that every element of the composite is of identical composition. More commonly, composite materials demonstrate clumped regions of densely packed fibres separated by resin lakes (Figure 4.3b), where there are gross differences between local regions of the composite. A realistic optimum for this class of composite material would be uniform random distribution of the fibres (Figure 4.3c).

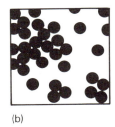

(a)   (b)   (c)

Figure 4.3 (a) Uniform hexagonal packing. (b) Non-uniform clumped distribution. (c) Uniform random packing

The fibre distribution in melt impregnated tape of carbon fibre reinforced PEEK approximates to this random packing (Figures 4.1 and 4.2), and is independent of processing history.

Of course the fibre distribution at a local scale depends on the scale at which it is inspected. With reinforcement fibres of the order $7\,\mu m$ diameter, if we inspect regions of $1\,\mu m \times 1\,\mu m$, it is possible, indeed probable, that we shall encounter 100% fibre or 100% resin. As the sample size increases, so the breadth of distribution narrows and, for a typical sample of CF/PEEK, Table 4.2 shows maximum, mean and minimum fibre fractions.

Table 4.2 Variation of fibre volume fraction with sample size

Sample scale	Area	Width of prepreg tape 125 μm thick	Fibre volume fraction		
			Maximum	Mean	Minimum
$3.3\,\mu m \times 3.3\,\mu m$	10 (μm^2)		1.00	0.61	0.00
$10\,\mu m \times 10\,\mu m$	100 (μm^2)		0.91	0.61	0.00
$33\,\mu m \times 33\,\mu m$	1000 (μm^2)	8 μm	0.80	0.61	0.25
$100\,\mu m \times 100\,\mu m$	10,000 (μm^2)	80 μm	0.73	0.61	0.40
$0.33\,mm \times 0.33\,mm$	0.1 $(mm)^2$	0.8 mm	0.68	0.61	0.50
$1\,mm \times 1\,mm$	1 $(mm)^2$	8 mm	0.65	0.61	0.58
$3.3\,mm \times 3.3\,mm$	10 $(mm)^2$	80 mm	0.63	0.61	0.59

In addition to heterogeneous distribution of fibres throughout the composite, there is one specific variation though the thickness of a laminate. A resin rich layer exists in between plies of preimpregnated tape. This is most easily visible in laminates with plies of different orientation (Figure 4.4).

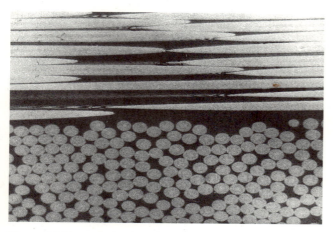

Figure 4.4 The resin-rich layer between plies of different orientation (fibre diameter 7 µm)

This resin rich region occurs because fibres in one layer cannot cross the boundary into plies of different orientation. As a first approximation, the typical interply resin rich layer is comparable to the thickness of the reinforcing fibre. This layer is therefore thicker for prepregs based on large diameter glass fibres than for fine carbon fibres, and represents a greater proportion of the total prepreg thickness in such systems. For a nominal 125 µm thick carbon fibre prepreg the two resin rich layers on either side will represent about 5% of the total thickness and about 13% of the total resin in the composite. The concentration of resin in the surface region of the plies means a corresponding depletion of resin in the interior; in that region of a prepreg of nominally 61% by volume the actual fibre packing is about 65% by volume. Non uniformity of the resin rich surface layers is reflected in a tendency for the prepreg to curl, with the concave surface having the excess of resin. This is due to the higher shrinkage of neat resin than composite material on cooling. It is the fusing of resin rich layers, sometimes formed during the processing operation, which assures the bonding between plies on which the integrity of the composite structure depends.

Some reports of comparative mechanical testing of competitive product forms include micrographs of resin fibre distribution[6,7]. These studies suggest that good properties accompany uniform fibre distribution, but other factors such as interface control or fibre attrition are also in play. There are many studies of the effects of fibre content on mechanical performance, but at this time I know of no controlled experimental investigation, or applied theoretical study, of the influence of fibre/matrix distribution. There have been practical demonstrations of the value to be gained from incorporating tough interply layers into brittle matrix thermosetting systems[8]. It is possible that a clumped distribution of fibres may give improved resistance to local microbuckling, and resin rich interlayers may reduce thermal stresses in materials, but, at this time, there is no study systematically varying the interply thickness layer in naturally tough thermoplastic systems. These all appear valuable areas for academic research. Until more evidence is available, pragmatism suggest that it is desirable to aim for as uniform a fibre distribution as possible in such systems, with a minimum resin rich surface consistent with good interply adhesion.

4.2.2 Fibre straightness

In comparison to the paucity of opinion about resin/fibre distribution, most people take for granted the desirability of straightness of the fibres. Composites based on wavy or off axis fibres are less stiff and strong, although it should be noted that nature, in general, eschews ultimately straight fibres, preferring slight misalignment as a way of achieving toughness in her composites[9]. A particular concern with wavy fibres is that they are envisaged as being weak in compression (Appendix 2). The carbon fibre tows from which prepregs are made usually contain 12,000 filaments per tow, and these tows are usually supplied wound on spools. Certain manufacturers deliberately supply twisted fibres because of benefits such products give to textile technologies. Alternatively, some fibre tows are deliberately entangled to open up the interior of the fibre bundle, thereby facilitating impregnation. Even in the case of nominally untwisted fibres, such as Hercules AS4, some slight twisting is inevitably introduced as the collimated tow is wound on to and off spools. There is also the issue that shaping processes may lead to distortion of the fibre orientation, but that aspect will be considered in Chapter 5. This section considers only the misalignment of fibres that is inherent within the composite material. Some small level of misalignment appears impossible to avoid, and we must consider both its magnitude and consequences.

Yurgatis[10] has developed a method of measuring fibre misalignment, and has applied his technique to the study of carbon fibre/PEEK. In his technique the composite is sliced and polished at an angle of about 5° to the principle fibre axis. Microscopic inspection shows the cut fibre surfaces as elongated ellipses from the aspect ratio of which the actual orientation of the fibre can be calculated with an accuracy of 0.25°. For melt preimpregnated carbon fibre/PEEK, Yurgatis reports that 83% of the fibres are within ±1° of being perfectly aligned, and all the fibres are within ±4°. The degree of fibre alignment is essentially the same both in the plane and out of the plane of the preimpregnated tape (Table 4.3).

Although these results indicate very high fibre collimation, the presence of a few angled fibres is significant. Remembering that all the fibres were initially of the same length, and initially straight, the presence of angled fibres indicates that those fibres are either discontinuous or stretched. If we assume that the fibres are stretched then the 3% of fibres that are at 2.5° off axis are strained by 0.1%, while the very occasional fibre that is 4° off axis is likely to be strained by 0.25% relative

Table 4.3 Fibre misalignment in preimpregnated tape[11]

Fibre	Fibre fraction	
misalignment	In plane	Out of plane
0 ± 0.25°	0.24	0.22
0.5° ± 0.25°	0.41	0.43
1.0° ± 0.25°	0.23	0.24
1.5° ± 0.25°	0.08	0.07
2.0° ± 0.25°	0.01	0.01
2.5 to 4°	0.02	0.03

to its colleagues. These are a very few fibres indeed, but the consequence of that strain can become manifest when the prepreg is moulded.

On moulding, Yurgartis notes a significant change in in-plane fibre orientation. There is a general flattening of the fibre orientation profile and typically 10% of the fibres can be about 3° to 4° off alignment. Further, it is noted that there is a tendency for adjacent fibres to misalign cooperatively, so that qualitatively the orientation pattern changes, as in Figure 4.5.

Figure 4.5 Changes in orientation pattern during moulding (exaggerated sketches)

Kanellopoulos and his colleagues[12], who had also observed similar effects with thermosetting composites, attribute this tendency of the fibres to kink to the relaxation of stresses in the matrix. I prefer the argument that this phenomenon is due to shrinkage of the few off axis, strained, fibres in the prepreg when the constraint of the matrix is removed by melting. As they shrink, those fibres provide a compression stress, and since the matrix is molten, there is nothing to stop the highly aligned fibres crimping to accommodate that shrinkage. This feature can be simply noted by melting a piece of prepreg in an unconstrained fashion, when the fibres may crimp out of the plane of the tape but each wrinkle will have a few taut fibres crossing it (Figure 4.6).

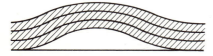

Figure 4.6 Out of plane crimping bridged by taut fibres (exaggerated sketch)

In this experiment it is sometimes noted that, within 20 mm of the free ends, no such crimping occurs. In this part of the moulding it may be assumed that the stretch in the few twisted fibres can relieve itself at the free ends. This slight fibre crimping can be, and indeed usually is, prevented by applying suitable constraint during the moulding process. Such constraint is virtually always a feature of any practical forming operation involving significant shaping.

We must characterize the level of fibre misalignment which can be induced as a result of this relaxation process (Table 4.4). Qualitatively similar results are seen for both uniaxial and quasi-isotropic samples.

Table 4.4 Maximum fibre misalignment[13]

	Prepreg (%)	Moulding (%)
0 to 1° off axis	83	36
1 to 2° off axis	14	32
2 to 3° off axis	3	22
3 to 4° off axis	very few	10

It is also necessary to study the effect of such fibre misalignment on the properties of the composite moulding. The most obvious property affected by fibre wrinkling is cosmetic appearance. The human eye is particularly sensitive at picking out regular wavy undulations. This is made even easier by a difference in reflectivity of fibre reinforced plastics dependent on the angle which light makes with the direction of the fibres. Thus, although few human eyes would readily detect variations of 4° from a control direction in a single fibre, a periodic cooperative wave with a maximum off axis angle of ±4° and a wavelength of about 6 mm is readily detectable as mottling. Because of the ability of the few twisted fibres to relieve their stress at the free edge of the moulding, such mottling is sometimes observed as a band in the centre of the moulding surrounded by apparently straight fibres at the edge: this again enhances visual discrimination. The presence of such a visual, cosmetic, effect naturally creates concern for the mechanical performance of the composite, especially with respect to compression performance.

We can study the effect of such fibre misalignment on properties by deliberately selecting mouldings, or sections of mouldings, with and without fibre wrinkling or by deliberately making wrinkled prepreg and constructing laminates therefrom. There are three ways to make such wrinkled prepreg. The first method is to stress relieve the prepreg out of line, so that the stretched fibres relax and crimp their colleagues. The second is deliberately to manufacture the prepreg tape without tension on the fibres. This tension contributes to the spreading of the fibre tows in the impregnation process, so that the absence of tension can result in other cosmetic problems, such as splitting of the tape web. The third option is to look at woven fabric materials, where the weaving process naturally introduces a rather larger level of cooperative crimp and gives rise to similar anxieties. The latter comparison can only be made with 0/90 laminates. Tables 4.5–4.7 compare properties of uniaxial, cross plied and quasi-isotropic laminates with and without fibre misalignment of the level indicated in Table 4.4.

Table 4.5 Effect of fibre misalignment on compression strength in uniaxial mouldings

	IITRI compression strength (MN/m^2)
Moulded from standard prepreg	
1 No perceptible misalignment	1,100 (100)*
2 With severe fibre wrinkling	880 (130)
Moulding from stress relieved prepreg	
3 With fibre misalignment	1,030 (130)
Moulding from low tension prepreg	
4 With fibre misalignment	1,030 (75)

* Standard deviation in parentheses.

The absolute level of fibre misalignment in 2,3 and 4 is similar, although its visibility is less obvious in the mouldings of 4, since there is less cooperative misalignment of the fibres to attract the eye. A qualitative cosmetic judgement of the fibre misalignment in the three prepreg tapes rates standard prepreg good, stress relieved prepreg gross, and low-tension prepreg as irregular.

Table 4.6 Effect of fibre misalignment on compression strength in cross plied laminates

	IITRI compression strength (MN/m^2)
Moulding from standard prepreg	
No perceptible misalignment	675 (50)*
Moulding from preimpregnated woven fabric[7]	
With controlled fibre wrinkling	700 (40)

* Standard deviation in parentheses.

Table 4.7 Comparison of compression strength of quasi-isotropic laminates

	IITRI compression strength (MN/m^2)
Mouldings from standard prepreg	
No perceptible misalignment	630 (25)*
With severe fibre wrinkling	625 (25)

* Standard deviation in parentheses.

Tables 4.5 to 4.7 indicate that the structural performance of multiaxial thermoplastic composite materials is not significantly affected by the level of fibre wrinkling associated with stress relief in the prepreg on moulding. The one area where a significant (20%) reduction in compression strength can be observed is for thick uniaxial stacks. In this case it appears that there may be cooperative buckling in adjacent layers of prepreg, which exaggerates the problem. If the laminae are 'precrimped', by stress relieving out of line, by prepregging under low tension, or by using a woven fabric, then such cooperative wrinkling does not occur. We may therefore argue that while cooperative wrinkling of the fibres is undesirable, there is no evidence that random misalignment significantly affects properties.

Figure 4.7 Fibre organizations (exaggerated sketches)

For all its cosmetic appearance such fibre wrinkling only reflects a change in mean fibre misalignment from 0.6° to 1.4°.

Despite the fact that random fibre misalignment resulting from stress relief in the prepreg has no deleterious effect on performance, it is desirable to control it in practice. This control is most readily achieved by maintaining the fibres under constraint during processing. This constraint is applied naturally during a shaping process, particularly when using a diaphragm forming process[14], since the shaping process tries to stretch the inextensible prepreg. Perversely, it is in the manufacture of simple flat laminates, such as are commonly used for coupon tests of mechanical performance, that it is least easy to constrain the fibres. The use of matched die moulds, which are also often preferred for making coupon test specimens, can further accentuate the level of fibre misalignment by inducing local transverse flow processes to iron out slight irregularities in the prepreg thickness. Mild steel tooling used to make such matched die moulds has a higher thermal expansion coefficient than that of carbon fibre composites, and shrinkage of such tooling during mould cooling can induce compressive stress into the composite and further contribute to fibre wrinkling. Very rapid cooling also tends to lead to exaggerated fibre wrinkling. Such rapid cooling is usually achieved by transferring the moulding from a hot press to a cold one. During such a transfer process the constraints on the fibres are released. To minimize such fibre wrinkling in matched die mouldings, the constraint should be maintained at all times. The tendency for test coupons to be more subject to fibre wrinkling then shaped components is advantageous. Any slight degradation in properties is likely to be more severe in the test coupon than will be found in practical mouldings: the natural constraints of shaping processes will mean that fibre misalignment within a ply will usually be less than that in the simple coupon specimens from which its properties are determined.

Most composite structures are based on multi-angle laminates. The change in orientation at each layer impedes the development of cooperative fibre wrinkling. The most likely cases where some deterioration should be anticipated are in thick, rapidly cooled, uniaxial layers. Such mouldings should be made with special care to maintain constraint on the fibres at all times. If that constraint cannot be maintained, then there is the option of precrimping the prepreg out of line. This provides a cosmetically unattractive product, where further tendency to crimp is eliminated. There is no cooperative crimping within the plies. The elimination of such cooperative crimping prevents the 20% reduction in compression strength that might otherwise occur in thick, fast cooled, uniaxial mouldings. For most composite mouldings there is no evidence of significant deterioration in properties associated with any fibre misalignment inherent in the preimpregnated tape product.

4.3 Morphology of the matrix

Whereas there is a scarcity of publications giving detailed descriptions of fibre distribution and its importance in determining performance, no such reticence is evident on the subject of crystalline texture in the resin. The surge of publications in this field was stimulated by the observation that the level of crystallinity, and its organization in terms of spherulitic structure, varied with processing history. There is an optimum processing window for such a system[15]. The existence of a processing window immediately stimulated a number of groups to carry out processing outside that range, and hence create sub-optimal morphologies in order to determine if the excellent properties of carbon fibre reinforced polyetheretherketone could be degraded by such history. Those studies, which have been reviewed by Corrigan[16], concluded that only minor changes result from such deviations.

One benefit of those studies has been a remarkably detailed understanding of the crystalline texture of PEEK resin in composites. In this context, a series of papers by Blundell and his colleagues[17-20] is particularly notable, especially when taken in conjunction with their papers on neat resin PEEK[21]. The crystallization behaviour in composite essentially follows that of neat resin, except that the triaxial constraint of the reinforcing fibres slightly slows down crystallization. The fibres also interact with the crystallization process and can strongly influence the spherulitic texture. The studies of melting and crystallization vary from pragmatic[22] to consideration of detailed molecular motion[23]. Apart from differences of detail on the interaction between the fibre and crystallization behaviour, a strong concensus has appeared.

4.3.1 The amorphous phase

Rapid cooling from the melt at cooling rates in excess of 15°C/s results in amorphous resin. Such cooling rates can be achieved on transferring the molten material into a cold press of high thermal mass. The fastest cooling rates known to have been recorded are ~100C/s[24]. Achieving amorphous resin requires quenching from the melt to below the glass transition temperature in less than 10 seconds. This can be achieved on thin mouldings, but heat transfer considerations mean that it is very difficult to make amorphous mouldings that are more than 1 mm thick. In the amorphous state PEEK resin has enhanced ductility but somewhat reduced solvent resistance. In the presence of a solvent amorphous PEEK will usually crystallize, so that it has an inbuilt protection mechanism. Mouldings that are made amorphous can readily be crystallized by an annealing process[25–31]: a recommended annealing cycle to develop optimum crystallinity is 60 seconds in the temperature range 220°–270°C.

Kenny[32] suggests that the amorphous phase can potentially be exploited to carry out some forming process just above Tg in the temperature range between 150°C and 200°C. In this temperature range the amorphous resin is highly viscous and no fusion processes should be expected, but some level of creep forming may be practicable to tailor shapes to a preferred geometry before finally 'setting' the structure by raising the temperature to crystallize it. In such 'rubber phase' processes, shaping deformations may result in orientation of the polymer matrix molecules and so induce some anisotropy into that phase. The presence of even very low levels of crystallinity severely inhibits flow.

Thermal ageing just below the glass transition temperature can reduce the free volume in amorphous polymers. Raising the temperature above Tg and subsequently quenching restores the maximum free volume. Low free volume increases stiffness and resistance to creep at the expense of some loss of toughness. Free volume effects are less commonly encountered with semi-crystalline matrix resins. In the case of carbon fibre reinforced polyetheretherketone 1,000 hours at 120°C produces no detectable effect on properties[33].

4.3.2 Crystallization and crystallinity

Two techniques are used for the study of crystallinity: WAXS (wide angle X-ray scanning)[34–37], from which detail of the crystal structure can be deduced (Figure 4.8); and DSC (differential scanning calorimetry)[38,39], which leads to a more

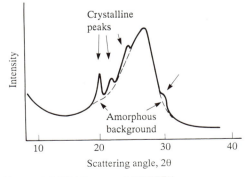

Figure 4.8 WAXS trace of CF/PEEK

qualitiative estimate but is especially valuable for studying the dynamics of crystallization and melting (Figures 4.9 and 4.10). The two techniques give satisfactory quantitative agreement for this class of composite material.

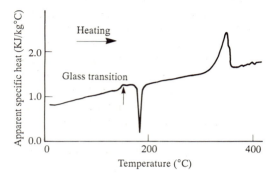

Figure 4.9 DSC trace for amorphous CF/PEEK

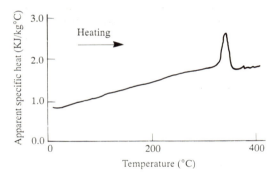

Figure 4.10 DSC trace for semi-crystalline CF/PEEK

The matrix in PEEK based composites crystallizes slightly more slowly than neat resin under the same conditions[40] (Table 4.8). Since pressure accelerates crystallization, this slowing down can be interpreted as due to the physical constraint of the fibres producing a hydrostatic tension on the matrix. When the

Table 4.8 Comparison of crystallization and melting behaviour of neat resin and matrix in composite by differential scanning calorimetry[41]

	Cooling*		Heating*	
	Crystallization temperature °C	Latent heat kJ/kg	Melting temperature °C	Latent heat kJ/kg
Resin	300	49	343	44
Composite after processing	294	14	342	12
Resin in composite		43		39

* Measured at heating and cooling rates of 0.3°C/sec.

fibres are present, natural growth of the spherulites is impeded to follow geometrically tortuous paths, again slowing down the crystallization process.

The crystallization of polyetheretherketone depends first on the state of the melt from which it is crystallized. It is preferred that this melt should be at a temperature of at least 360°C, so that predetermined nucleii associated with residual traces of unmelted crystallites are eliminated[42]. Shukla and Sichina[43] reported that, in an early sample of carbon fibre/PEEK, APC-1, the crystallization of the matrix slowed down after time in the melt state, however, Sichina and Gill[44,45] demonstrated that, in the final optimized version of this material, APC-2, the crystallization behaviour is not affected by such exposure. Saliba[46] extended his study to hold times of 2 hours at 400°C with similar conclusions. Ma[47] has further demonstrated that repeated melting and crystallization does not significantly alter the crystallization behaviour of PEEK matrix composites. Provided the melt has been established within the recommended processing range of 360°C to 400°C the crystallization behaviour of PEEK matrix composites can be accurately defined.

The rate at which the melt is cooled to below the glass transition temperature is the primary processing variable that controls the level of crystallinity. This phenomenon has been extensively studied experimentally[48-51], and a consensus of those results is presented in Figure 4.11. Those results are consistent with studies of the isothermal rate of crystallization[53-56] (Figure 4.12).

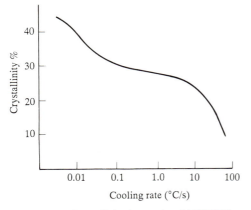

Figure 4.11 Crystallinity in matrix of CF/PEEK as a function of cooling rate, based on the review by Corrigan[52]

Kitano and Seferis[58] indicate that sample vibration can slow down the crystallization rate in carbon fibre/PEEK composites.

This strong body of experimental studies has produced an even stronger response from the modelling fraternity[59-68]. Although details of theory are hotly disputed, all authorities concur that the rate of crystallization and level of crystallinity can be accurately predicted. In addition to the observation of crystallinity, Cebe[69] detects a 'disordered', or paracrystalline, region, particularly associated with fast cooling cycles. For the recommended processing window, a cooling rate between 0.1 and

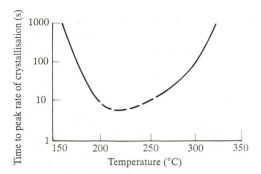

Figure 4.12 Time of peak rate of crystallization of CF/PEEK at constant temperature[57]

10°C/s, the optimum level of crystallinity in carbon fibre reinforced polyetheretherketone is between 25 and 35%.

The dependence of crystallinity on cooling rate naturally causes concern about crystallinity gradients within a moulding. That anxiety has been dispelled by Blundell and Willmouth[70] who showed that, during normal processing, although the centre of a moulding naturally solidifies after the surfaces, the cooling rate experienced through the critical temperature region of crystallization is virtually independent of position in the moulding. Only in extreme cases, when cooling thick mouldings into cold tools, was it possible to achieve amorphous surfaces and crystalline central regions. In such cases a short annealing cycle at 220°C will crystallize the surface layer.

The difficulty of obtaining amorphous surfaces, usually an undesirable result, is associated with the relatively high thermal conductivity of carbon fibre composites compared to normal plastics. It would not necessarily be true for composites based on other fibres, such as glass.

Although the matrix phase is essentially isotropic, some slight orientation of the primary growth axis of the crystals is detected along the fibre direction[71]. This orientation arises because of the constraints imposed by the dimensionally stable fibres upon the matrix as it cools and shrinks.

We have noted how amorphous materials may be crystallized by a subsequent annealing process. Annealing processes can also be used to increase the crystallinity of normally crystallized materials[72–76]. Such processes have been used particularly to examine the 'overcrystallized' state of PEEK composites. Berglund[77] reports achieving 43% crystallinity after 50 hours at 310°C. The use of an annealing process usually produces a variation in the crystal melting process observed by differential scanning calorimetry. Annealed samples usually have an additional melting process at a slightly higher temperature than that at which they were annealed[78]. Annealing just below the melting point can raise the melting point by a few degrees. In fast crystallizing matrices, such as polyetheretherketone, annealing is seldom required in order to achieve optimum crystallinity in the moulding, but adventitious annealing may occur, for example as the result of a high temperature bonding process.

High crystallinity leads to enhanced stiffness and reduction in toughness. The possible effect on toughness, the single most outstanding service property of thermoplastic composite materials, led a number of investigators to examine the possibility of significantly affecting this property by processing outside the recommended range and obtaining excessively high crystallinity[79-91]. Typical results from those studies are summarized below.

There have been a few studies on the effect of cooling rate and crystallinity on stiffness and strength. In comparison with materials crystallized in the preferred cooling rate range, Sichina and Gill[92] show that rapidly cooled samples, having lower crystallinity, have a more clearly defined loss process at the glass transition temperature. They report room temperature shear moduli of quenched samples to be about 10% lower than standard. Talbott[93] reports a 30% reduction in shear modulus of quenched samples. Grossman and Amateau[94] report a 20% reduction in in plane shear modulus for quenched samples and observe a 10% higher shear modulus for very slow cooled samples. The latter authors also explored the axial and transverse modulus and strength over the cooling rate range 0.01 to 40°C/s. Within the preferred processing range 0.1 to 10°C/s. these values appear to be independent of cooling rate. Table 4.9 compares the values at the two extremes with that mean.

Table 4.9 Axial, shear and transverse stiffness and strength in comparison with standard processing[95]

	Axial modulus	Axial strength	Shear moduli	Transverse modulus	Transverse strength
Slow cooled, excessive crystallization	1.0	1.0	1.1	1.0	0.9
Quenched, incomplete crystallization	0.9	1.1	0.9	0.9	1.0

Grossman and Amateau[96] also compare their mechanical property data with the manufacturer's data sheet values and, in the recommended processing range, they obtain good agreement.

Measurements of fracture toughness[97-99] as a function of cooling rate, as collated by Corrigan[100], are shown in Table 4.10. These results suggest a variation of about ±20% about a mean, with significant variation due to testing technique. As expected, the less crystalline materials are tougher.

Table 4.10 Fracture toughness as a function of cooling rate[101]

Cooling rate °C/sec.	G_{IC} (KJ/m^2)	
0.01	1.5 (±0.5)	overcrystallized
0.1	1.6 (±0.5)	⎫
1.0	1.7 (±0.5)	⎬ recommended processing range
10	1.8 (±0.3)	⎭
100	2.0 (±0.2)	incomplete crystallization

Leicy and Hogg[102] have studied the impact properties of standard processed and quenched mouldings of carbon fibre/PEEK. They observe that the incompletely crystallized samples absorb 10% less energy at punch through, but that they are significantly less prone to delamination on impact.

The resistance to delamination is of particular significant in compression testing after an impact event. Work carried out at the Boeing Company[103] studied post impact compression behaviour of laminates as a function of processing history. These results, in comparison with standard processed materials, are shown in Table 4.11. Although significant gains in toughness could be achieved by using incompletely crystallized mouldings, there is only a small loss in this critical property associated with excessive crystallization.

Table 4.11 Compression strength after impact as a function of level of crystallinity in comparison with optimally processed samples[104]

Thermal history	Compression strength after impact (relative to standard processing)
Annealed at 320°C, excess crystallinity	0.9
Quenched, incompletely crystalline	1.2

Damage tolerance is also an issue where the damage is designed into the structure, as is the case with the inclusion of a hole. Table 4.12 compares open hole tension and compression data at different levels of crystallinity with standard processed materials[105].

Table 4.12 Open hole properties relative to standard processed samples[106]

Thermal history	Open hole tensile strength	Open hole compression strength
Annealed at 320°C, excessive crystallinity	1.0	1.0
Quench cooled, incomplete crystallization	1.05	1.0

The tougher matrix gives slightly enhanced open hole tensile strength. There is no loss of open hole compression performance for either extreme of processing.

Although excess crystallization reduces the fracture toughness, and associated properties such as damage tolerance, by about 10%, this barely erodes the tenfold advantage in toughness possessed by this class of material in comparison to conventional thermosetting composites. When excessive crystallization is present there is evidence of some slight gain in properties such as compression performance, which are dependent on the matrix stiffness (Appendix 2). Considering the extremes which have been attempted in order to produce a morphology which compromises the service performance of carbon fibre reinforced polyetheretherketone, and the ill success of such attempts, the manufacturers may be open to censure for being too conservative in their recommendations on processing. It is generally agreed that any practically achievable crystalline morphology of Aromatic Polymer Composite (APC-2) provides excellent composite material.

The most distinctive feature of the crystalline texture of polyetheretherketone is the spherulitic structure. Spherulites are families of crystallites radiating from a single nucleation point. Two issues of spherulitic structure are considered to be especially significant: potential weakness at interspherulitic boundaries, and 'on fibre' nucleation. Because spherulites grow outwards from a point there comes a time when two or more spherulites impinge. Since the resin is shrinking as it crystallizes, such impingement points represent regions of intense competition for a finite resource of matrix so that voids could result. Such voids could represent critical stress concentrations from which cracks might initiate. Any problems of interspherulitic weakness would be more likely in large spherulites[107,108], and large spherulites are usually associated with slower cooling rates[109], leading to a preference for fast cooling. Anxieties about interspherulitic weakness in Aromatic Polymer Composites are dispelled by the work of Crick and his colleagues[110], who have examined the failure surfaces of broken samples and shown that crack propagation is not associated with spherulite boundaries. Anomalous spherulite structure can sometimes be found when the matrix has been incompletely melted. That structure is associated with unmolten crystalline residues seeding a proliferation of small spherulites[111]. Dramatic early pictures[112] illustrated an influence of fibres on the nucleation of spherulitic structure when a single fibre was embedded in the resin. Theories of transcrystallization were expressed wherein the matrix resin crystallized on to the fibre surface, turning that interface into a continuum. So important was this considered to be that there were expressions of the desirability of destroying all the nuclei in the matrix phase so that the crystallization was preferentially seeded at the fibre surface[113]. That argument leads to a preference for long times at high melt temperature. Preferential nucleation does indeed occur naturally at the surfaces of highly graphitized fibre[114]. Such nucleation can best be observed by a technique pioneered by Peacock[115], whereby a sample is cooled to a temperature at which it will crystallize and then, before crystallization is complete, the sample is quench-cooled. When such a sample is cut, polished and etched the growth of the spherulites can be carefully traced (Figure 2.7).

The success of this technique is, of itself, dramatic testimony to the way in which the rate and level of crystallization can be predicted. Attractive though theories of transcrystallization are, careful investigation by Blundell[116] has demonstrated that, while nucleation can occur either randomly within the matrix or at the fibre surface, the predominant site of spherulite nucleation in the standard grade of Aromatic Polymer Composite based on high strength carbon fibres is where two fibres touch, or come very close together (Figure 4.13).

The reason for this preferred growth site is presumably that the shrinkage stresses as the matrix cools and solidifies will be exaggerated at this point; stress is a

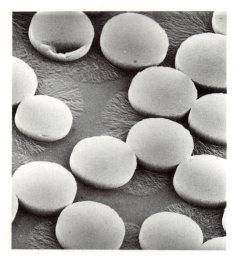

Figure 4.13 Spherulite growing from point where two fibres nearly touch

well-known nucleant for crystallization. This observation helps to explain the early reports of intense nucleation round single fibres in the resin[117,118]. Those observations were made on single fibres embedded in resin between glass slides and pressed out thinly to the thickness of the fibre. Where the fibre made contact with the glass slide, there was a similar, exaggerated, geometry to fibre–fibre contact (Figure 4.14).

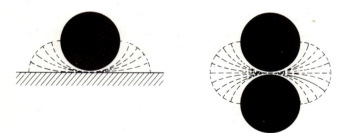

Figure 4.14 (a) Fibre and flat surface. (b) Fibre–fibre contact

The tendency for spherulites to be nucleated at the fibre surface is not of critical importance to the performance of composite materials. It is, however, verification of the molecular scale proximity of resin and fibre and so an elegant testimony to the quality of the impregnation process to achieve fibre wetting. Dramatic though the spherulite structure of carbon fibre/PEEK composite is, it is only of secondary importance to the level of crystallinity in determining properties.

4.4 Internal stresses in the composite

Internal stresses in composite materials are often associated with the difference in thermal expansion of their components, the organization of those ingredients, and the processing history of the composite. Although not introduced as a designed element of the microstructure, internal stress contributes directly to the performance of the structure.

In a simple uniaxial carbon fibre composite the thermal expansion of the fibres is nearly zero while that of the resin phase is large and positive. As the composite is cooled from its processing temperature the resin tries to shrink, but that shrinkage is resisted by the stiff, inextensible fibres. This places the resin in tension and the fibres in compression. Using X-ray diffraction, Hank[119], determined such stresses in a carbon fibre/PEEK laminate as about $28 MN/m^2$. This agrees well with the calculation of Young[120], who deduced a figure of $34 MN/m^2$. Young observed that this would place the fibres under a compressive strain of 0.01% but he and Galiotis[121] used Raman spectroscopy to measure the apparent compressive strains in fibres in a preimpregnated tape at about 0.25%; however, the spectral data included interference from the PEEK matrix, and more recent measurements taking this into account have indicated compressive strains of the same order as the calculated values. Although stresses of the order $30 MN/m^2$ build up as the sample cools down, only a very small proportion of this stress level would be present while the composite is in a molten state. Manson and Seferis[122] have suggested that such stresses may contribute to buckling of the fibres during the cooling process: this is likely to be only a small factor, there being more potent forces at work (Section 4.2.2. above). The tensile stress on the resin resulting from this difference in thermal expansion is a significant factor, being about 25% of the matrix yield stress.

As the resin-fibre distribution in a composite is non-uniform, so also will be the internal stresses[123]. Since the maximum internal stress is most likely to be the cause of any problem, minimizing that stress is desirable. This leads to a preference for a uniform fibre distribution in the composite. In the case of the Aromatic Polymer Composite family, heterogeneity of resin–fibre distribution is controlled, and minimized, at the preimpregnation stage.

In addition to the constraint imposed by the fibres on the resin, the whole of the composite is placed under constraint by the solidification taking place as a result of cooling. Cooling takes place from the outside inwards. The surface layers of a moulding are thus solidified first. They will be placed in compression by the subsequent shrinkage of the internal core and will themselves represent a tensile constraint on that part of the moulding. Manson and Seferis[124] used an elegant method of demonstrating this effect by the use of release plies between each layer of his moulding. They were then able to study the deformation in each layer independently. For carbon fibre reinforced polyetheretherketone they showed an approximately parabolic stress profile, with a maximum compressive stress of about $40 MN/m^2$ on the surface and a maximum tensile stress in the interior of the moulding of about $18 MN/m^2$. Chapman[125] has provided a detailed theoretical account of this mechanism. This stress applies to the composite as a whole and not just the components. Thus in the surface layers of a moulding, during rapid cooling, a compressive stress of 40MN/m² on the composite can balance the tensile stress on the matrix, owing to the difference in thermal expansion of the ingredients.

Uniaxial mouldings are rarely used in practice; angle plied laminates are more common. All those ingredients to internal stress noted above will also apply to angle plied laminates, but there is extra constraint imposed by the lay up. This constraint is most clearly demonstrated by taking a two ply 0/90 laminate, moulding it and cooling it in a flat press and then inspecting it. Instead of a flat plate, the panel will be curved (Figure 4.15). Further, that curvature is bistable. Bending it

Figure 4.15 Unbalanced 0/90 laminate

flat will cause it to flip over to the opposite curvature. This results from the differential thermal expansion of the composite along and across the fibre direction. Along the fibres this property is dominated by the stiff fibre and so virtually zero; across the fibre direction it is of similar magnitude to the thermal expansion of the matrix. Nairn and Zoller[126], Zoller[127] and Jeronimidis and Parkyn[128] have used this bimetallic strip technique to study internal stresses and their effects in carbon fibre/PEEK and other composites. Table 4.13 shows a comparison of predicted and observed curvatures for different laminates.

Table 4.13 Radius of curvature of cross-plied laminates[129]

Laminate	Thickness (mm)	Radius of curvature (mm) Predicted by bimetallic strip	Measured experimentally
(0/90)	0.25	50	46
(0₂/90₂)	0.5	100	105
(0₄/90₄)	1.0	201	200
(0₂/90₂)₂	1.0	823	850

The studies of Nairn and Zoller[130] and Jeronimidis and Parkyn[131] are self-consistent. Jeronimidis and Parkyn indicate that, although there are large volume changes between the melting point and the glass transition temperature, three quarters of the total stress build up occurs between Tg and ambient temperature. For (90₂/0₄/90₂) symmetric laminates, wherein the stress cannot be released by curvature, Jeronimidis and Parkyn calculate a compressive stress along the fibre direction of 42 MN/m² and a tensile stress across the fibre direction of

$42\,MN/m^2$. That tensile stress is approximately half the transverse tensile strength of the composite material. Jeronimidis and Parkyn were able to demonstrate that first ply failure, in the transverse plies, happened under a nominal additional stress in the transverse layer of $38\,MN/m^2$. This gave a total transverse tensile failure stress of $80MN/m^2$, in agreement with their expectation. Dramatic though the effects of internal stress in cross plied laminates can be, they can be readily predicted.

One major area of concern associated with internal stress is the possibility of microcracking. Nairn and Zoller[132], Zoller[133] and Coxon[134] have studied this effect for a range of composites. The internal stress levels in CF/PEEK are higher than those with the other thermoplastic and thermosetting resins studied by Nairn and Zoller, but CF/PEEK is reported to be much less susceptible to microcracking. Nairn and Zoller attribute this to the toughness of PEEK resin, but Coxon suggests that the uniformity of fibre/resin distribution is also a significant factor. The resistance of thermoplastic composites to microcracking depends both on the ingredients and how they are integrated.

Microcracking can be a severe problem in cross plied PEEK composites based on fragile, ultra high modulus, fibres. Barnes[135] shows that such microcracking is due to transverse failure of the fibres. Inspection of micrographs of such transverse ply cracking suggests that the microcracking allows a crack of the order $1.5\,\mu m$ to form approximately every $250\,\mu m$ along the ply, indicating strain release of the order 0.6%. The transverse modulus of ultra high modulus carbon fibre/PEEK is of the order $9\,GN/m^2$, and this suggests that the internal stress release is of the order $33\,MN/m^2$, consistent with the calculations of Jeronimidis and Parkyn[136].

Cross ply lamination, besides inducing a tensile stress in the transverse plies, also induces a compressive stress in the fibre direction. Consider a thick quasi-isotropic moulding built up of individual uniaxial layers. The coefficient of thermal expansion, CTE, is small but still six times that of a uniaxial ply[137]. As the moulding cools from the melt, the shrinkage of the quasi-isotropic plies will place individual plies into compression (Table 4.14).

Table 4.14 Axial compression strain induced in a uniaxial ply during solidification of a quasi-isotropic laminate

Temperature range	CTE (per °C) QI	CTE (per °C) UNI	CTE (per °C) QI–UNI	Compression strain
343 to 143°C	6.0×10^{-6}	1.0×10^{-6}	5.0×10^{-6}	1.0×10^{-3}
143 to 23°C	2.9×10^{-6}	0.5×10^{-6}	2.4×10^{-1}	0.3×10^{-3}
			Total	1.3×10^{-3}

In this table the compressive strain in the uniaxial laminates is calculated as the temperature gradient multiplied by the difference in CTE. Note that in this case the full temperature range must be considered, because, unlike the resin, where most of the stress builds up below Tg, the fibre has no way of relieving the compressive stress build up. The total compressive strain induced is of the order 0.13%. The existence of this strain is sometimes seen when samples have been subjected to high velocity impact, or have been incorrectly processed, so that there is a delamination at the surface ply. That delamination appears as a blister (Figure 4.16).

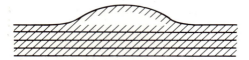

Figure 4.16 Delamination of a surface ply (exaggerated sketch)

If a gross approximation to such delamination is considered to be of a triangular shape, the release of 0.13% compressive strain in the surface ply would allow a blister height approximately 1 mm for every 40 mm delamination. This is qualitatively consistent with practical observation. This is a significant compression strain, approximately 15% of the compressive strain to failure in a uniaxial laminate.

Up to this point all our attention has been directed to flat laminates. Most practical mouldings are shaped, and we must also consider thermal stresses and deformations associated with the difference in the low thermal expansion in the plane of the moulding (fibre dominated) and the high value through its thickness (resin dominated). Spencer[138] shows that, as a moulding cools and becomes thinner, the radius of curvature tightens (Figure 4.17).

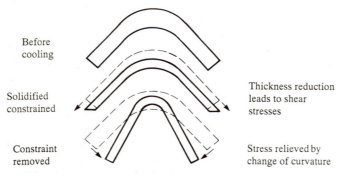

Before cooling

Solidified constrained

Constraint removed

Thickness reduction leads to shear stresses

Stress relieved by change of curvature

Figure 4.17 Influence of differences in thermal expansion through the thickness and in the plane of the moulding (exaggerated sketches)

This phenomenon is sometimes called 'spring forward' or 'tow in' on moulding. Typically a 90° bend will spring forward as a result of this thermally induced stress by about 2°. Zahlan and O'Neill[139] have extended the rigorous analysis for single curvature structures[140] by the use of finite element modelling to achieve capacity for predicting such distortions in double curvature structures.

This section has identified four factors contributing to internal stress in a composite structure: the difference in thermal expansion of the matrix and reinforcement, temperature gradients through the thickness of the laminate, the lay up geometry, and the shape of the moulding. These factors are not unique to carbon fibre/PEEK or indeed to thermoplastic composites in general, but, because of the high processing temperatures, the thermal stresses and strains are higher in such composites than in most other systems. Despite that magnitude, independent observers[141–144] conclude that the commercially available preimpregnated tape form of carbon fibre/PEEK composites are remarkably resistant to the microcracking defects which might reasonably be expected. This resistance is testimony to the toughness of the resin, the strength of the interface, and the uniform distribution of fibres in the preimpregnated tape.

4.5 Quality in composites: the control of microstructure

Ultimately we seek the optimum service performance from composite materials. Chapter 2 considered the properties of the ingredients which ultimately determine what properties we can aspire to. Chapter 3 explored how those ingredients could be integrated, and the level of the ultimate performance that could be achieved. In this chapter we have considered the microstructure of the composite, the reproducibility of which determines the level of confidence that we can place on the performance of the product. The microstructure of all composite materials is complex. Thermoplastics in general, and carbon fibre/PEEK in particular, are no exception to this rule. Complexity leads to anxiety, which was manifest in an explosion of papers studying the crystallization behaviour of the matrix of PEEK composites around 1983 to 1986. In this area a consensus has been arrived at and, although details of theory are hotly debated, good predictive capacity is available. Similarly, there has been scientific investigation of the internal stresses in laminates, such that their magnitude and influence can be foretold. Only in respect of the resin–fibre distribution is there a dearth of detailed study, unless one is to include the frequent laconic reports that the fibres are well distributed and there are no significant voids. The absence of comment otherwise indicates that all that can be achieved is being achieved. Thus, where quality assurance is based on knowledge and understanding of microstructure, for thermoplastic composites, exemplified by preimpregnated carbon fibre reinforced polyetheretherketone, we probably know more about the system than we do of any other composite material in service today. To turn that knowledge to account we must next consider how to make structures from those preimpregnated tapes.

References

4–1   J. A. BARNES, N. J. CLAYDEN, P. JACKSON, T. A. CARPENTER, L. D. HALL AND P. JEZZARD, 'Defect Detection in Carbon Fibre Composite Structures by Magnetic Resonance Imaging', *J. Materials Science Letters* **9**, 1165–1168 (1990).

4–2   J. A. E. MANSON AND J. C. SEFERIS, 'Void Characterisation Techniques for Advanced Semicrystalline Thermoplastic Composites', *Science of Engineering Composite Materials* (December 1988).

4–3   *Ibid.*

4–4   E. SILVERMAN AND R. J. JONES, 'Property and Processing Performance of Graphite/PEEK Prepreg Tapes and Fabrics', *SAMPE Journal,* **24**, 4, pp. 33–42 (1988).

4–5   *Ibid.*

4–6   *Ibid.*

4–7   U. MEASURIA AND F. N. COGSWELL, 'Aromatic Polymer Composites: Broadening the Range', *SAMPE Journal,* **21**, 5, pp. 26–31 (1985).

4–8   R. B. KRIEGER, K. HIRSCHBUEHLER AND R. E. POLITI, 'Thermoset Interleaved Resin Matrix Composites with Improved Compression Properties', EPO 01 33 280 (1985).

4–9   G. JERONIMIDIS AND J. E. GORDON, 'Composite Materials with High Work to Fracture', *Phil. Trans Royal Soc. of London.* A294, pp. 545–550 (1980).

4–10   S. W. YURGATIS, 'Measurement of Fiber Misalignment in Aligned Fiber Composites and the Influence of Lamination on Misalignment Distribution', *Proc. Am. Soc. for Composites 3rd Tech Conf.,* pp. 587–596, Technomic Publishing (1988).

4–11   *Ibid.*

4–12   V. N. KANELLOPOULOUS, B. YATES, G. H. WOSTENHOLM, M. I. DARBY, J. EASTHAM AND D. ROSTRON, 'Fabrication Characteristics of a Carbon Fibre-Reinforced Thermoplastic Resin', *J. Materials Science Letters,* **24**, pp. 4000–4003 (1989)

4–13   See 4–10.

4–14   A. J. BARNES AND J. B. CATTANACH, 'Advances in Thermoplastic Composite Fabrication Technology', Materials Engineering Conference, Leeds (1985).

4–15   F. N. COGSWELL, 'Microstructure and Properties of Thermoplastic Aromatic Polymer Composites', 28th National SAMPE Symposium, pp. 528–534 (1983).

4–16   E. CORRIGAN, D. C. LEACH AND T. MCDANIELS, 'The Influence of Processing Conditions on the Properties of PEEK Matrix Composites', in *Materials and Processing – Move in the 90's, Proc. 10th Int. Eur. Chapter Conf. of SAMPE,* Materials Science Monographs, **55**, pp. 121–132, Elsevier (1989).

4–17   D. J. BLUNDELL, J. M. CHALMERS, M. W. MACKENZIE AND W. F. GASKIN, 'Crystalline Morphology of the Matrix of PEEK–Carbon Fibre Aromatic Polymer Composites, Part 1: Assessment of Crystallinity', *SAMPE Quarterly,* **16**, 4, pp. 22–30 (1985).

4–18   D. J. BLUNDELL AND B. N. OSBORN, 'Crystalline Morphology of the Matrix of PEEK–Carbon Fibre Aromatic Polymer Composites, Part 2: Crystallization Behaviour', *SAMPE Quarterly,* **17**, 1, pp. 1–6 (1985).

4–19   D. J. BLUNDELL AND F. M. WILLMOUTH, 'Crystalline Morphology of the Matrix of PEEK–Carbon Fibre Aromatic Polymer Composites, Part 3: Prediction of Cooling Rates During Processing', *SAMPE Quarterly,* **17**, 2, pp. 50–58 (1986).

4–20   D. J. BLUNDELL, R. A. CRICK, B. FIFE, J. A. PEACOCK, A. KELLER AND A. WADDON, 'Spherulitic Morphology of the Matrix of Thermoplastic PEEK/Carbon Fibre Aromatic Polymer Composites', *Journal of Materials Science,* **24**, pp. 2057–2064. (1989)

4–21   D. J. BLUNDELL AND B. N. OSBORN, 'The Morphology of Poly(aryl ether ether ketone)', *Polymer,* **24**, pp. 953–958 (1983).

4–22   J. D. MUZZY, D. M. ABOUELNAER AND A. O. KAYS, 'Melting and Solidification of APC-2', 33rd International SAMPE Symposium, pp. 1319–1330 (1988).

4–23   H. X. NGUYEN AND H. ISHIDA, 'Molecular Analysis of the Melting and Crystallization Behaviour of Poly(aryl-ether-ether-ketone)', Case Western Reserve University (1985), *Polymer Composites,* **8**, pp. 57–73 (1987).

4–24   A. DUTHIE, private communication (1984).

4–25   See 4–15.

4–26   W. J. SICHINA AND P. S. GILL, *Characterization of PEEK/Carbon Fibre Composites by Thermal Analysis,* Dupont Instrument Company, Delaware (1986).

4–27   M. QI, X. XU, J. ZHENG, W. WANG AND Z. QI, 'Isothermal Cyrstallisation Behaviour of Poly(ether-ether-ketone) (PEEK) and its Carbon Fibre Composites', *Thermochimica Acta,* **134**, pp. 223–230 (1988).

4–28   L. BERGLUND, *The Effect of Annealing on the Delamination of Carbon Fibre/PEEK Composite,* Linkoping University (1986).

4–29   G. M. K. OSTBERG AND J. C. SEFERIS, 'Annealing Effects on the Crystallinity of Polyetheretherketone (PEEK) and its Carbon Composites', *J. Appl. Polym. Sci.,* **33**, p. 29 (1987).

4–30   P. CEBE, 'Annealing Study of Poly (etheretherketone)', *J. Materials Science* (in press, 1989).

4–31   J. KENNY, A. D'AMORE, L. NICOLAIS, M. IANNONE AND B SCATTEIA, 'Processing of Amorphous PEEK and Amorphous PEEK based Composites', *SAMPE Journal*, **25**, 4, pp. 27–34 (1989).

4–32   *Ibid.*

4–33   F. N. COGSWELL, 'The Processing Science of Thermoplastic Structural Composites', *International Polymer Processing*, **1**, 4, pp. 157–165 (1987).

4–34   See 4–17.

4–35   M.-F. SHEU, J.-H. LIN, W.-L. CHUNG AND C.-L. ONG, 'The Measurement of Crystallinity in Advanced Thermoplastics', 33rd International SAMPE Symposium pp. 1307–1318 (1988).

4–36   D. E. SPAHR AND J. M. SCHULTZ, 'Determination of the Matrix Crystallinity of Composites by X-ray Diffraction', submitted to *Polymer Composites* (1989).

4–37   P. CEBE, L. LOWRY AND S. CHUNG, 'Use of Scattering Methods for characterisation of Morphology in Semi-Crystalline Thermoplastics', SPE ANTEC (1989).

4–38   See 4–26.

4–39   C. C. M. MA AND S. W. YUR, 'Parameters Affecting the Crystallization of Carbon Fiber Reinforced Polyether Ether Ketone Composites', SPE 47th ANTEC pp. 1422–1429 (1989).

4–40   See 4–33.

4–41   See 4–33.

4–42   See 4–20.

4–43   J. G. SHUKLA AND W. J. SICHINA, 'Thermal Behaviour of Carbon Fibre Reinforced Polyetheretherketone', SPE ANTEC, pp. 265–267 (1984).

4–44   See 4–26.

4–45   W. J. SICHINA AND P. S. GILL, 'Characterization of Composites by Thermal Analysis', 31st International SAMPE Symposium, pp. 1104–1112 (1986).

4–46   T. E. SALIBA, D. P. ANDERSON AND R. A. SERVAIS, 'Process Modelling of Heat Transfer and Crystallization in Complex Shapes Thermoplastic Composites', *Journal of Thermoplastic Composite Materials*, **2** (1989).

4–47   C.-C. MA, J.-T. HU, W.-L. LIU, H.-C. HSIA, B.-Y. SHIEH AND R.-S. LUI, 'The Thermal, Rheological and Morphological Properties of Polyphenyle Sulphide and Polyetheretherketone Resins and Composites', 31st Int. SAMPE Symposium, pp. 420–433 (1986).

4–48   See 4–17.

4–49   See 4–18.

4–50   W. I. LEE, M. F. TALBOTT, G. S. SPRINGER AND L. A. BERGLUND, 'Effects of Cooling Rate on the Crystallinity and Mechanical Properties of Thermoplastic Composites', First Technical Conference of Proceedings of the American Society for Composites held in Dayton, Ohio 7–9 October 1986, pp. 119–128 (1986).

4–51   S. GROSSMAN AND M. AMATEAU, 'The Effect of Processing in a Graphite Fibre/ Polyetheretherketone Thermoplastic Composite', 33rd International SAMPE Symposium, pp. 681–692 (1988).

4–52   See 4–16.

4–53   See 4–18.

4–54   See 4–26.

4–55   See 4–27.

4–56   See 4–31.

4–57   See 4–18.

4–58   A. KITANO AND J. C. SEFERIS, 'Mechanical Influences on Crystallization in PEEK Matrix/Carbon Fibre Reinforced Composites', 35th International SAMPE Symposium, pp. 2272–2279 (1990).

4–59   See 4–39.

4–60   See 4–46.

4–61   C. N. VELISARIS AND J. C. SEFERIS, 'Crystallization Kinetics of Polyetheretherketone (PEEK) Matrices', *Polymer Engineering and Science*, **26**, pp. 1574–1581 (1986).

4–62   J. F. CARPENTER, 'Thermal Analysis and Crystallization Kinetics of High Temperature Thermoplastics', *SAMPE Journal*, **24**, 1, pp. 36–39 (1988).

4–63   P. CEBE, 'Application of the Parallel Avrami Model to Crystallization in PEEK', *Polymer Engineering of Science*, **28**, pp. 1192–1197 (1988).

4–64   A. M. MAFFEZZOLI, J. M. KENNY AND L. NICOLAIS, 'Modelling of Thermal and Crystallization Behaviour of the Processing of Thermoplastic Matrix Composites', in *Materials and Processing – Move into the 90's, Proc. 10th Int. Eur. Chapter Conf. of SAMPE*, Materials Science Monographs, **55**, pp. 133–144 Elsevier (1989).

4–65   J. C. SEFERIS, 'Crystallization Kinetics of PEEK Matrix Composites', Fifth Annual Meeting Polymer Processing Society, Kyoto, p. 179 (1989).

4–66   P. CEBE, 'Non-isothermal Crystallization of Poly(etheretherketone) Aromatic Polymer Composite', *Polymer Composites*, **9**, 4, pp. 271–279 (1988).

4–67   C. N. VELISARIS AND J. C. SEFERIS, 'Heat Transfer Effects on the Processing-Structure Relationships of Polyetheretherketone (PEEK) Based Composites', *Polymer Engineering and Science*, **28**, 9, pp. 583–591 (1988).

4–68   J. C. SEFERIS AND C. N. VELISARIS, 'Modeling-Processing-Structure Relationships of Polyetheretherketone (PEEK) Based Composites', 31st National SAMPE Symposium, pp. 1236–1245 (1986).

4–69   P. CEBE, L. LOWRY, S. Y. CHUNG, A. YAVROUIAN AND A. GUPTA, 'Wide-Angle X-Ray Scattering Study of Heat-Treated PEEK and PEEK Composite', *J. of Appl. Polymer Sci.*, **34**, pp. 2273–2283 (1987).

4–70   See 4–19.

4–71   See 4–20.

4–72   See 4–28.

4–73   See 4–29.

4–74   See 4–30.

4–75   J. C. SEFERIS, C. AHLSTROM AND S. H. DILLMAN, 'Cooling Rate and Annealing as Process Parameters for Semi-Crystalline Thermoplastic-Based Composites', SPE ANTEC 1467 (1987).

4–76   J. MIJOVIC AND T. C. GSELL, 'Calorimetric Study of Polyetheretherketone (PEEK) and its Carbon Fibre Composite', *SAMPE Quarterly*, **21**, 2, pp. 42–46 (1990).

4–77   See 4–28.

4–78   See 4–29.

4–79   See 4–16.

4–80   See 4–28.

4–81   See 4–50.

4–82   See 4–51.

4–83   M. F. TALBOTT, G. S. SPRINGER AND L. A. BERGLUND, 'The Effects of Crystallinity on the Mechanical Properties of PEEK Polymer and Graphite Fibre Reinforced PEEK', *Journal of Composite Materials*, **21**, 11, pp. 1056–1082 (1987).

4–84   P. T. CURTIS, P. DAVIES, I. K. PARTRIDGE AND J. P. SAINTY, 'Cooling Rate Effects in PEEK and Carbon Fibre-PEEK Composites', *Proc. ICCM VI and ECCM 2*, **4**, 401–412 (1987).

4–85   P. LEICY AND P. J. HOGG, '*The Effect of Crystallinity on the Impact Properties of Advanced Thermoplastic Composites*', ECCM3, pp. 809–815 Elsevier (1989).

4–86   C.-C. M. MA, S. W. YUR, C. L. ONG AND F. W. SHEU, 'Effect of Annealing on the Properties of Carbon Fiber Reinforced PEEK Composite, 34th International SAMPE Symposium, pp. 350–361 (1989).

4–87   W. SICHINA AND P. S. GILL, 'Characterization of Composites using Dynamic Mechanical Analysis', 33rd International SAMPE Symposium, pp. 1–11 (1988).

4–88   R. J. DOWNS AND L. A. BERGLUND, '*Apparatus for Preparing Thermoplastic Composites*', Linkoping Institute of Technology (1986).

4–89   P. DAVIES, W. CANTWELL, H. RICHARD, C. MOULIN AND H. H. KAUSCH, '*Interlaminar Fracture Testing of Carbon Fibre/PEEK Composites Validity and Application*', ECCM 3, pp. 747–755, Elsevier (1989).

4–90   P. VAUTEY, 'Cooling Rate Effects on the Mechanical Properties of a Semi-Crystalline Thermoplastic Composite', *SAMPE Quarterly*, **21**, 2, pp. 23–28 (1990).

4–91   P. DAVIES, W. J. CANTWELL, P.-Y. JAR, H. RICHARD AND H. H. KAUSCH, 'Cooling Rate Effects in Carbon Fibre/PEEK Composites', 3rd ASTM Symposium on Composite Materials, Orlando (1989).

4–92   See 4–26.

4–93   See 4–83.

4–94   See 4–51.

4–95   See 4–51.

4–96   See 4–51.

4–97   See 4–83.

4–98   See 4–84.

4–99   Y. LEE AND R. S. PORTER, 'Crystallization of PEEK in Carbon Fibre Composites', *Polymer Engineering and Science*, **26**, 9, pp. 633–639 (1986).

4–100 See 4–16.

4–101 See 4–16.

4–102 See 4–85.

4–103 NATIONAL MATERIALS ADVISORY BOARD, 'The Place for Thermoplastic Composites in Structural Components', National Research Council, NMAB-434 (1987).

4–104 Ibid.

4–105 Ibid.

4–106 Ibid.

4–107 See 4–15.

4–108 See 4–43.

4–109 See 4–15.

4–110 R. A. CRICK, D. C. LEACH, P. J. MEAKIN AND D. R. MOORE, 'Fracture and Fracture Morphology of Aromatic Polymer Composites', Journal of Materials Science, 22, pp. 2094–2104 (1987).

4–111 See 4–20.

4–112 See 4–15.

4–113 See 4–99.

4–114 J. A. PEACOCK, B. FIFE, E. NIELD AND C. Y. BARLOW, 'A Fibre-Matrix Interface Study of Some Experimental PEEK/Carbon Fibre Composites', First International Conference on Composite Interfaces, Cleveland (1986).

4–115 J. A. PEACOCK, B. FIFE, E. NIELD AND R. A. CRICK, 'Examination of the Morphology of Aromatic Polymer Composite (APC-2) using an Etching Technique', presented at the first International Conference on Composite Interface Cleveland, 27–30 May 1986.

4–116 See 4–20.

4–117 See 4–15.

4–118 See 4–43.

4–119 V. HANK, A. TROOST AND D. LEY, 'Use of X-Ray Diffraction ot Measure Lattice Strain and Determine Stress in Carbon Fiber Reinforced PEEK', Kunststoffe-German Plastics, 78, 11, pp. 41–43 (1988).

4–120 R. J. YOUNG, R. J. DAY, M. ZAKIKHARIC AND I. M. ROBINSON, 'Fibre Deformation and Residual Thermal Stresses in Carbon Fibre Reinforced PEEK', Composites Science and Technology, 34, pp. 243–258 (1989).

4–121 C. GALIOTIS, N. MELANITIS, D. N. BATCHEDLDER, I. M. ROBINSON AND J. A. PEACOCK, 'Residual Strain Mapping in Carbon Fibre/PEEK Composites', Composites, 19, 4, pp. 321–324 (1988).

4–122 J. A. MANSON AND J. C. SEFERIS, 'Internal Stress Determination by Process Simulated Laminates', SPE 45th Annual Technical Conference and Exhibition, pp. 1446–1449 (1987).

4–123 See 4–33.

4–124 See 4–122.

4–125 T. J. CHAPMAN, J. W. GILLESPIE, JR, R. B. PIPES, J. A. E. MANSON AND J. C. SEFERIS, 'Prediction of Process-Induced Residual Stresses in Thermoplastic Composites', submitted to Journal of Composite Materials (1989).

4–126 J. A. NAIRN AND P. ZOLLER, 'Residual Thermal Stresses in Semicrystalline Thermoplastic Matrix Composites', Fifth International Conference on Composite Materials ICCM-V, pp. 931–946 (1985).

4–127 P. ZOLLER, 'Solidification Processes in Thermoplastics and their Role in the Development of Internal Stresses in Composites', Proc. Am. Soc for Composites 3rd Technical Conf., pp. 439–448, Technomic Publishing (1988).

4–128 G. JERONIMIDIS AND A. T. PARKYN, 'Residual stresses in carbon fibre thermoplastic matrix laminates', Journal of Composite Materials, 22, pp. 401–415 (1988).

4–129 Ibid.

4–130 See 4–126.

4–131 See 4–128.

4–132 See 4–126.

4–133 See 4–127.

4–134 B. R. COXON, J. C. SEFERIS AND L. B. ILCEWICZ, 'The Effects of Process Variables on Transverse Matrix Cracking in High Performance Composites', Process Manufacturing International, 4, pp. 129–140 (1988).

4–135 J. A. BARNES AND F. N. COGSWELL, 'Thermoplastics for Space', SAMPE Quarterly, 20, 3, pp. 22–27 (1989).

4–136 See 4–128.

4–137 See 4–33.
4–138 J. M. O'NEILL, T. G. ROGERS AND A. J. M SPENCER, 'Thermally Induced Distortion in the Moulding of Laminated Channel Sections', *Mathematical Engineering in Industry,* **2**, 1, pp. 65–72 (1988).
4–139 N. ZAHLAN AND J. M. O'NEILL, 'Design and Fabrication of Composite Components: Spring-Forward Phenomenon', *Composites,* **20**, 1, pp. 77–81 (1989).
4–140 See 4–138.
4–141 See 4–126.
4–142 See 4–127.
4–143 See 4–128.
4–144 See 4–134.

5 Processing science and manufacturing technology

Processing converts the integrated product form into its designed shape. This chapter seeks: to define the fundamental operations in this stage; to understand how they can be controlled; to investigate how they can be technologically exploited; and to consider the implications of manufacturing on the service performance of the composite material. The service performance of a composite structure calls for three interactions: material, design and processing, as illustrated in Figure 5.1.

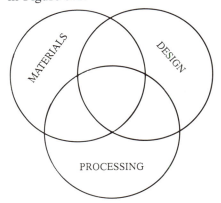

Figure 5.1 The critical interactions to achieve service performance

The availability of product form (Chapter 3) may constrain design, or the design may dictate the suitability of different product forms. Geometrical shape and product economics, determined at the design stage, may limit the range of manufacturing technologies from which choice can be made, while the availability of processing technology may constrain design or dictate product form. Processing of the material may alter fibre organization and morphology. The ultimate service performance of a structure will reflect the extent to which the inherent properties of the product form and design are translated by the conversion processes into an end product.

The manufacturing process is central to this translation. Despite this central position, it occupies a time span usually measured in minutes, or hours at most, in the life history of a thermoplastic material. Structures are likely to be designed for a service life of twenty years, and the passing phase of processing is sometimes ignored, with the assumption that if you take a little longer, press a little harder, or tailor the raw material product form to a specific application, then everything will be all right. This approach continues to be workable for the bespoke manufacture of small numbers of components. It is, however, an attitude that is inappropriate for the breakthrough we seek into mass production processing. Where mass production technology is the target for translating composite materials from aerospace technology into the field of general industrial products, the processing stage must be central to the study.

5.1 Processing science

The ultimate objectives of studying processing science are threefold: to be able to predict how a material will behave in a known manufacturing operation; to learn how to tailor a material product form so that it processes better; and to devise novel conversion processes to achieve designed end products. There are two approaches to these goals. Analysts[1-4] start by trying to isolate relevant elements of the process and to measure the response of the materials: they then seek to synthesize a process model. Modellers[5-9] prefer to define the process dynamics; by using that model in conjunction with experiments they then deduce the apparent characteristics of the material. Ideally these two approaches should meet on common ground. At this time (1990) the conclusions can differ significantly, and there is a considerable amount of work to be done in order to rationalize these different views and arrive at a consensus. Until that consensus is achieved, we shall not have true predictive capability but must judge the approaches by how their predictions accord with empirical observation.

Being a protagonist in the 'analytical' camp, I am not qualified to present the arguments for the 'modelling' school objectively. In following the analytical approach I shall attempt to indicate where discrepancies between the two approaches appear to arise, in the hope that future work can elucidate these points.

The analytical approach to processing must separately address three areas of materials science: chemical change, thermophysical properties, and deformation characterization.

5.1.1 Chemical change

In the study of processing operations nothing is so unrewarding as to create a model system and then discover that the properties of the material have been changing throughout the process. It is desirable to establish at an early stage if, and how, such changes may occur, so that they can be avoided wherever practicable. In thermoplastic processing there are three common factors which may lead to chemical change: heat, oxygen and water[10].

Thermoplastic structural composites are processed at high temperature, and thermal stability sets an upper limit to the processing window of materials like

PEEK. Groves[11] and Scobbo[12] have demonstrated that there is no significant change in the melt viscosity of CF/PEEK for 1 hour under normal processing conditions. Besides melt viscosity, another property that is sensitive to any changes in chemical structure is crystallization behaviour. Sichina and Gill[13] have shown that the crystallization behaviour of the optimized CF/PEEK prepreg does not change with normal melt history, and Ma[14] has shown that there are no significant changes in crystallization behaviour even after repeated processing. Direct measurements of the molecular weight of polymer extracted from such a composite[1] indicates that after 2 hours at 400°C there is a slight (10%) increase in molecular weight. This result indicates that any changes that are taking place during normal processing are, if anything, tending to enhance the toughness of the polymer matrix. Two hours is a longer time than would be anticipated for processing with fully impregnated thermoplastic composite materials, and we may safely conclude that, for preimpregnated carbon fibre/PEEK, processing within the manufacturer's recommended temperature range, 360–400°C, leads to no deleterious effect.

Excellent thermal stability is a feature of the polyketone family produced by nucleophilic polymerization (see Chapter 2). Carpenter[15] shows no significant weight loss in carbon fibre/PEEK preimpregnated tape products until 550°C. Similar polymers produced by other routes may be less stable.

Processing of 'Victrex' PEEK is possible at higher temperatures provided that the time scale is reduced and oxygen is excluded (Table 5.1).

Table 5.1 Suggested maximum time scales as a function of processing temperature for preimpregnated CF/PEEK

Temperature	Suggested maximum times at temperature
400°C	2 hours
450°C	15 minutes
500°C	2 minutes
550°C	15 seconds
600°C	2 seconds

It is possible to use flash heating such as flame, laser or hot gas in some rapid processing technologies without adversely affecting the properties of the composite.

In some thermoplastic composite materials significant chain extension during processing may actually be a sought after feature and an integral part of the material response[16]. Where this is the case, it is desirable to appreciate how the viscosity and crystallization behaviour will change during processing. The maximum predictive capability and optimum quality assurance will be achieved where such changes are minimized.

The effects of heat are aggravated by the presence of oxygen. With PEEK resin, oxygen tends to produce crosslinking, and so an increase in viscosity and a reduction in ease of crystallizing. Oxygen, from the air, first attacks any exposed surface of the material. If preimpregnated CF/PEEK tape is melted in air for about an hour, the surfaces will become so viscous that they cannot be fused together. The issue of oxidation is of particular importance in those product forms, such as commingled fibres or powder impregnated materials, which have a large surface to volume ratio. In particular, where the surface of the carbon fibre is exposed, there is the issue of forming the interface in the presence of an active agent to be considered. Where the product form has a large surface to volume ratio, inert atmosphere or vacuum processing is strongly recommended in order to achieve optimum control. This issue is less critical in the case of the boardy preimpregnated product forms, where there is only approximately one twentieth of the surface area exposed (Table 5.2), but, even so, Cattanach[17] strongly recommends vacuum wherever possible.

Table 5.2 Surface area of different product forms

Product form	Surface area/millilitre
Preimpregnated tape	$\approx 0.02 \, m^2$
Commingled fibres	$\approx 0.40 \, m^2$

In the context of providing an inert atmosphere it should be noted that 'Victrex' PEEK, prepared by the nucleophilic reaction process, contains residual traces of diphenylsulphone which is used as a reaction medium in the polymerization process. At melt temperature this sublimes and, in an enclosed system, will displace the air from surfaces to be fused, providing an intrinsic inert atmosphere during processing before being reabsorbed into the polymer. At normal service temperature it is present as a solid. The level of residual diphenylsulphone to provide this function is usually less than 0.1%. Very much larger amounts can be included without leading to any adverse effect on properties[18], although excess diphenylsulphone may lead to a condensate forming on cold parts of the processing equipment.

Many commercial engineering polymers are formed by condensation reactions. Such polymers can be prone to degradation when melted if they contain any water. Water molecules tend to react with the chain, causing scission. In such hydrophilic polymers water content must be very closely controlled. Other polymers, such as polyethersulphone and polyetherimide, will absorb significant quantities of water, which will normally volatilize during processing. Such volatilization processes can result in a scarred surface on the moulding or blisters in the structure. Polyetheretherketone resin is strongly resistant to water pick up. It has an equilibrium water content of about 0.25%, achievable after prolonged exposure to boiling water. Although such saturation has no detectable effects on the properties of carbon fibre/PEEK composites, it can give rise to cosmetic defects during processing. If water is present, then drying is strongly recommended (Table 5.3).

Table 5.3 Recommended drying conditions for preimpregnated CF/PEEK

Temperature	Time
150°C in air	2 hours
150°C in vacuum	30 minutes
60°C in air	24 hours

Water is a more serious problem with glass fibre reinforced plastics, where the water may react with the surface of the fibre. With such composites, drying before processing is always recommended. Any need for predrying of material can be reduced if suitable vacuum conditions are applied as part of the processing cycle.

The commercial grades of preimpregnated CF/PEEK based on polymer made by nucleophilic polymerization have excellent chemical stability. This allows a broad processing window to be established, in which the physical properties of the material pertaining to the processing operation can be accurately defined.

5.1.2 Thermophysical properties

In any thermoplastic polymer heat exchange, and the thermodynamic properties associated with such temperature changes, play an important role in defining process economics. Often it is the time taken to solidify a formed component that will be the dominant factor in determining the rate at which components can be made. Uneven temperature profiles resulting from process dynamics can be a cause of non-uniform deformation or failure to conform to shape in the desired manner. They can also cause internal stress in the composite (Chapter 4). For structural composite materials such as carbon fibre/PEEK the processing temperatures are high, and so the associated thermophysical effects are large, but only as large as would be expected.

Polymers are good insulating materials: carbon fibres are moderate conductors of heat. Some, ultra high modulus, carbon fibres are exceptionally good conductors of heat, and such fibres are sometimes selected where the composite material is to have a heat management, as well as structural, role. Working values of the heat transfer characteristics of carbon fibre/PEEK composite materials based on 60% by volume of high strength carbon fibres, as they relate to processing, are noted in Table 5.4. Such properties have been characterized[19-21] and Appendix 8 includes a more complete description throughout the temperature range of service as well as processing.

Table 5.4 Working values of heat content and heat transfer of PEEK/AS4

	Values
Heat content at 400°C relative to 20°C	540 kJ/kg
Specific heat at 400°C	1.7 kJ/kg °C
Coefficient of thermal diffusion (20°C to 400°C):	
across the fibre direction	0.3×10^{-6} m²/s
along the fibre direction	2.9×10^{-6} m²/s

It is important to note the anisotropy of thermal conductivity associated with the anisotropic nature of the reinforcing fibres. Usually heat transfer across the fibre direction – through the thickness of the moulding – is the dominant property of interest, but axial heat transfer can also be important, especially in continuous processes such as filament winding, tape laying, roll forming and pultrusion. Because of the insulating nature of the matrix resin, which is amplified if voids are present, and the slightly twisted and tangled nature of the heat conducting fibres (Chapter 4), through thickness heat exchange may well include local conduction along the fibres. This can lead to non-uniform heat flow through the material. Such non-uniformity is minimized in systems that have a good uniform distribution of well wetted fibres.

As well as the heat transfer characteristics within the composite material, it is also desirable to consider the surface heat transfer coefficient with any contacting medium (Table 5.5). This is particularly significant in the case of thin sections.

Table 5.5 Surface heat transfer coefficients[22]

Contacting surface	Surface heat transfer coefficient
Water	$\approx$ 2000 W/m²°C
Metal	$\approx$ 400 W/m²°C
Air at 10 m/s	$\approx$ 50 W/m²°C
Still air	$\approx$ 10 W/m²°C

As a 'black body', a carbon fibre composite is especially suitable for radiative heat transfer.

From such a solid basis of understanding, excellent thermal modelling capability has been demonstrated[23-30]. This includes heat exchange of rates up to 100°C/s[31]. Blundell and Willmouth[32] demonstrated that, during normal processing of carbon fibre/PEEK, although the surface layers of the moulding cool more rapidly than the centre, the heat transfer rate in the temperature range of relevance to crystallization is essentially the same throughout the moulding.

Subsequent experimental work[33] confirms this conclusion. As a result, we may assert with confidence that, during normal processing, the crystalline structure

(Chapter 4) will be similar throughout the whole thickness of the moulding, so that it can be inspected non-destructively, using surface X-ray reflectance techniques[34]. This important result arises from the relatively high thermal diffusivity within the composite, combined with the constraint on heat transfer at the surface of the moulding. The fact that surface layers of reduced crystallinity can be achieved under laboratory conditions, using especially designed moulds[35], serves only to emphasize the breadth of this part of the processing window.

Besides its effect on crystallization behaviour, the thermal gradient in a moulding can affect internal stress. This has been discussed in detail in Chapter 4.

In any heat exchange process there will be dimensional change. In carbon fibre composites axial expansion is constrained by the low thermal expansion coefficient of the fibre. Across the fibre direction the thermal expansion is more closely related to that of the resin. These properties have been characterized by Barnes[36] and detailed values are included in Appendix 8. Table 5.6 shows typical working values for carbon fibre/PEEK composites based on 61% by volume of high strength carbon fibre.

Table 5.6 Linear thermal expansion of carbon fibre/PEEK composites

	Solid state 23 to 143°C per °C	Rubbery region 143 to 343°C per °C
Uniaxial moulding:		
along fibres (0°)	0.4×10^{-6}	(0)
transverse (90°)	30×10^{-6}	(80×10^{-6})
Quasi-isotropic moulding:		
in plane (−45°, 90°, 45°, 0)s	2.9×10^{-6}	(7×10^{-6})
through thickness	58×10^{-6}	(130×10^{-6})

Estimated values in parentheses.

The low thermal expansion of the moulding is beneficial in giving accurate dimensional control, but note that the difference between in plane and through thickness properties can also cause distortion of shaped mouldings, as discussed in Chapter 4. The low thermal expansion of the composite material can be embarrassing when using matched metal tooling if that tooling receives the same thermal history as the composite. Such metal tools usually have high thermal expansion and will shrink on to the moulding, causing difficulties of extraction unless this factor has been considered at the design stage. Olsen[37] prefers to use graphite tooling to achieve optimum mouldings. Note, however, that when transferring hot composite into a cold metal tool, a preferred technology for mass production, the thermal excursion of the tooling is small by comparison to that of the composite being moulded: the thermal expansion mismatch is less evident. When the requirement is for precision structures, it is necessary to take detailed account of the thermal expansion of both the composite and the mould.

The integrated function of thermal expansion coefficient is the density function. The density of carbon fibre/PEEK composite based on 61% by volume carbon fibres decreases from $1,600 \, kg/m^3$ at 20°C, to $1,583 \, kg/m^3$ at the glass transition temperature, $1,550 \, kg/m^3$ at 300°C and $1,530 \, kg/m^3$ in the melt (Appendix 8). There is a small change in density at the cystalline melting point during heat up of the moulding. The temperature at which this occurs during cooling depends upon the temperature of crystallization, and so on the cooling rate: this effect is discussed in detail in Chapter 4.

During normal processing the crystallization temperature will be about 250°C. The total density change during a processing cycle is about 4%: virtually all this change is in the thickness of the moulding. Accommodating such shrinkages can provide problems when through thickness shrinkage is constrained by shape factors. For example, in thick tubular structures cooled uniformly from the melt the wall will attempt to reduce in thickness, but both the inner and outer circumferences, and so diameters, do not. The stresses so generated can cause delamination, and the thermal history and external constraints in such processes must be designed with care. Much of this volume change occurs in the melt or during crystallization, where the matrix is able to flow; most of the remainder of the shrinkage occurs in the rubbery region above the glass transition temperature, where relaxation can occur to relieve the build-up of stress.

5.1.3 Rheology

In continuous fibre reinforced plastics deformation is constrained by the high loading of continuous collimated fibres. The inextensibility of these fibres effectively means that tensile deformations along the fibre axis are forbidden. To understand how to mould double curvature structures from inextensible material is a non-trivial exercise and requires an appreciation of those deformations processes which are allowable. These are summarized in Figure 5.2. They include resin percolation through and along the fibre bed, transverse fibre flow, shearing of

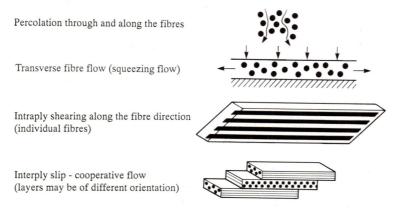

Percolation through and along the fibres

Transverse fibre flow (squeezing flow)

Intraply shearing along the fibre direction (individual fibres)

Interply slip - cooperative flow (layers may be of different orientation)

Figure 5.2 Flow mechanisms in preimpregnated products

individual fibres past one another along the fibre direction, and lamina slip of plies past one another.

In addition to these flow mechanisms, in continuous collimated products it is possible to incorporate additional flow, a superficial stretching, by replacing the continuous fibres by discontinuous aligned fibres[38] or crimped or twisted fibres[39]. While such product forms allow stretch to occur during moulding, the local flow processes (Appendix 9) remain dominated by shearing.

Percolation of resin through the reinforcing fibres is a healing flow. It allows local redistribution of resin and, in particular, the formation of a resin rich interlayer that permits bonding of different ply layers. The bed of reinforcing fibres is anisotropic: there can be resin percolation both along and across the fibres (Figure 5.3).

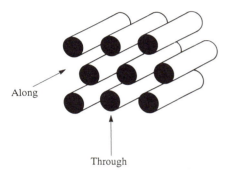

Along

Through

Figure 5.3 Anisotropic permeability

The description of such percolation can be made in terms of the D'Arcy Permeability Coefficient[40].

$$\text{The D'Arcy Permeability Coefficient} = \frac{\text{viscosity} \times \text{mean velocity} \times \text{thickness}}{\text{pressure drop}}$$

There are several analytical and experimental studies of resin permeability in fibre beds[40–46]. Analytical studies[47,48] indicate that the permeability along the fibre direction is significantly greater than that through the thickness. The permeability is higher with large diameter fibres: larger fibres, larger holes, easier flow. The permeability decreases by approximately one order of magnitude as fibre volume fraction increases from 55% to 65% by volume. Wheeler[49] notes that transverse permeability is not significantly affected by local fibre organization at those high volume fractions. Lam and Kardos[50] have made direct measurements of axial and transverse permeability of carbon fibres, using viscous oils. Those results are in satisfactory agreement with the analytical predictions (Table 5.7).

The permeability through a stack of plies of different orientation is reduced in comparison with that through a uniaxial laminate. Chapter 4 noted the more clearly defined resin rich layers in cross plied laminates. Forming that resin rich layer

Table 5.7 Permeability of carbon fibres by viscous resins (carbon fibre diameter 7 μm).

D'Arcy Permeability Coefficient (m^2)

Volume Fraction %	Analytical solution[51]		Experimental results[52]	
	Axial	Transverse	Axial	Transverse
50	4.5×10^{-13}	1.8×10^{-13}	5.8×10^{-13}	1.4×10^{-13}
55	3.5×10^{-13}	1.0×10^{-13}	3.1×10^{-13}	0.9×10^{-13}
60	2.3×10^{-13}	0.5×10^{-13}	1.9×10^{-13}	0.5×10^{-13}
65	1.5×10^{-13}	0.2×10^{-13}	1.0×10^{-13}	0.3×10^{-13}

concentrates the fibres more densely in the centre of the ply. That more dense array is of lower permeability. Lam and Kardos[53] indicate that, if the plies are at an angle of 45°, permeability is reduced by 25%, and for cross-plied laminates the reduction is 50%.

Very much higher values of permeability have been reported by other workers[54]. However, these appear to be measured by means of air fluidization techniques, where it may be difficult to prevent opening of the fibrous network in the turbulent air stream.

The permeability coefficient is determined by the geometric constraint of the fibre bed. The ease with which thermoplastic can flow through that constraint, and many other aspects of processing rheology, are determined by the melt viscosity of the resin phase. Practical experience suggests that a melt viscosity in the range 10 to 10,000 Ns/m^2 is preferred: too high a viscosity makes it difficult to achieve fusion between prepreg layers, while too low a viscosity tends to create difficulty in preventing excess resin being squeezed out of the material. The optimization of viscosity must also consider other factors. Viscosity is particularly sensitive to molecular weight[55] which in turn controls toughness. Thus the service requirements may limit viscosity selection to the high viscosity range. Wetting out the fibres is made more difficult by the selection of high viscosity resins. The need is for well impregnated, tough matrix, composites: the available viscosity range can be severely constrained. For polymer melts, viscosity depends on temperature and shear stress. Typical melt viscosity data for PEEK resins are presented in Figure 5.4. Usually composite materials are based on low viscosity systems.

As well as dependence on temperature and shear stress, viscosity is also affected by pressure and flow geometry[56]. In the case of impregnation processes (Chapter 3) it has been suggested that high shear processes may be helpful in order to reduce the viscosity to achieve full wetting of the fibres. High pressure also has many advocates. Where fibre wetting is an integral part of the forming process, a very full understanding of the polymer rheology may be required. In the case of processing with preimpregnated feedstock the flow processes are usually at low stress and carried out under modest pressure. Under these circumstances the compressibility of the melt, the non-Newtonian effects, and the flow geometry aspects are of secondary importance. In the case of carbon fibre/PEEK composites processing, a typical working value of 300 Ns/m^2 may be used as the polymer melt viscosity.

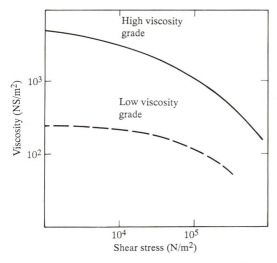

Figure 5.4 Viscosity characteristic of typical PEEK resins at 380°C; for viscosity at 360°C or 400°C multiply or divide viscosity at fixed shear stress by 1.35

At rest the polymer molecule occupies an entangled random coil configuration. During flow this configuration is disturbed. The recovery from such disturbance is characterized by the natural time of the melt[57]. This is defined as the ratio of melt viscosity to melt elastic modulus. For PEEK melts the elastic modulus is about $3,000 \, N/m^2$ [58], so that, for a melt viscosity of $300 \, Ns/m^2$, the relaxation time is about 0.1 sec. If the duration of deformation is longer than this time scale, then the flow will be essentially viscous; if shorter, the material will have a strong memory of its former state. Similar mechanisms will determine the interdiffusion of polymer chains across a boundary and, as a first approximation, I suggest that full adhesion across such a boundary will be achieved in about five relaxation times. This suggests a time scale for self diffusion of two melt fronts in contact as about half a second. Like resin percolation, which enables resin surfaces to come into contact, the self diffusion process is a major healing flow, which helps to determine the quality of the laminate.

The fibrous network also has deformation characteristics. Ideally the rod like reinforcements all lie parallel: in practice they are slightly twisted and intermingled, so that individual fibres are slightly bent. This bending of the fibres makes the fibre bed elastic and, in the absence of resin, it would fluff up slightly. Once the fibres are impregnated with resin, this tendency to spring apart is suppressed, but, as Gutowski[59] indicates, the elasticity of the fibre network cannot be ignored, especially when considering consolidation processes. The work of Barnes[60] suggests that for fully impregnated CF/PEEK composites, of 61% by volume fibre, a pressure constraint of 0.25 atmospheres is sufficient to overcome this elasticity of the fibre network and prevent void formation in the material. The twisting of the inextensible fibres also impedes certain deformation processes, including transverse flow[61] and axial intraply shear, whereby individual fibres are required to move past one another. It has been suggested[62] that the constraints imposed by the fibre network are responsible for the marked non-linear rheometric response of this class of composite material, but, at this time (1990), no definitive methods of characterizing the fibre bed, independent of the composite material, have been suggested.

When considering the rheology of a composite material as applied to processing science, the first question that must be asked is 'Is the rheology measured under laboratory conditions the same as pertains in processing?'[63]. There are two factors to be considered: the scales of the deformation history in respect of strain amplitude and process time. In respect of strain amplitude most composite processes cause strains of less than one unit of shear. Thus, if we consider the simple bending of a beam composed of inextensible composite through a right angle (Figure 5.5), the magnitude of the shear strain can be readily shown to be two

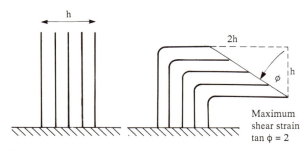

Figure 5.5 Bending of an inextensible beam

units of shear. A right angle is a not uncommon deformation requirement, but most shaping processes will lead to smaller deformations. There will, of course, be certain occasions where a small part of the material is subjected to much larger deformations, but, in terms of composite processing, 10 units of shear would be a large deformation indeed. This scale of deformation, 0–2 units of shear strain, with extension to 10 units of strain, is one at which it is both convenient and conventional to measure rheological response in the laboratory. In respect of time scales we are concerned with processes which may take less than 1 second in respect of rapid stamping to more than 1 hour in the case of autoclave processing. Here the major concern is for the thermal stability of the material. Groves[64] demonstrates that, for fully impregnated CF/PEEK composites, there should be no significant changes in rheology within such a time scale. From these considerations it is apparent that normal laboratory studies of the rheology of thermoplastic composites are relevant to the actual processing behaviour.

The questions remain: what rheological responses? And how should they be measured?. In a system wherein extension is forbidden by the inextensible fibres deformation must rely on shear processes. Two modes of intraply shearing, wherein individual fibres move relative to one another, can be distinguished: axial and transverse intraply shear (Figure 5.6).

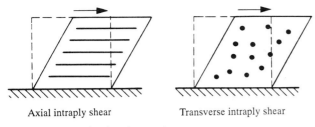

Axial intraply shear   Transverse intraply shear

Figure 5.6 Intraply shearing modes

At small amplitude of deformation these can be conveniently studied in oscillatory flow[65-67]. Conventional oscillatory rheometers operate in a torsional mode and, for composite materials based on continuous collimated fibres, such torsional experiments include both axial and transverse shear modes of deformation (Figure 5.7).

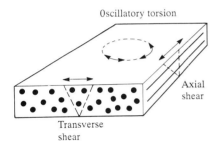

Figure 5.7 Torsional deformation

Rogers[68] has developed a method by which the individual components of such deformations can be derived and Groves[69] and Cathey[70] have applied such theories to deduce the axial and transverse shear deformation behaviour. Scobbo and Nakajima[71] have constructed an oscillatory sliding plate rheometer to measure those individual responses directly. Those studies[72-74], summarized in Appendix 10, conclude that the axial intraply shearing viscosity at low amplitude is about 30% higher than that in the transverse mode, and has a similar dependence on strain amplitude up to one unit of shear. All the experimental studies[75-79] demonstrate a markedly non-linear response, with indications of a yield stress of about 10^3 N/m^2. The interpretation of dynamic viscoelastic response in oscillatory shear depends upon the model used. My preference[80] is to use the Maxwell model instead of the more commonly used Voigt elements. By using this interpretation and translating to equivalent steady flow response, by defining maximum shear rate as angular velocity multiplied by strain amplitude, the oscillatory flow results can be substantially simplified[81] and presented in a form that is directly relevant to processing calculations. Since there appears to be only a small difference between axial and transverse intraply shearing, and since most practical deformations are also a compound of both these modes, the representation of torsional intraply shearing appears the most useful engineering data. Such data for carbon fibre/PEEK composite, representing apparent intraply shear viscosity as a function of shear stress, shear rate and strain amplitude, are shown in Figure 5.8 for a temperature of 380°C.

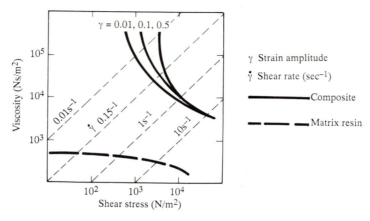

Figure 5.8 Apparent torsional intraply shear viscosity for carbon fibre/PEEK containing 61% volume carbon fibre[82]

Groves[83] provides comparative data for axial and transverse shearing modes (Figure 5.9). Those results show both a lower viscosity and a lower apparent yield stress for the transverse shearing mode.

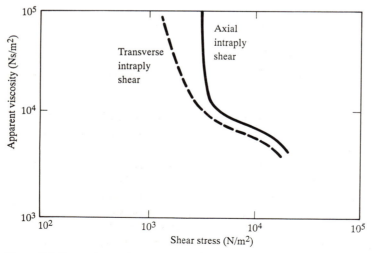

Figure 5.9 Comparison of axial and transverse shearing modes of CF/PEEK composite[84]

The temperature dependence of melt viscosity for the composite material is the same as that of the resin: a 10°C increase in temperature reduces viscosity by 17%. Such rheological measurements are readily accessible through conventional laboratory rheometers, such as the Rheometrics Mechanical Spectrometer.

Large amplitude, transverse intraply shear deformation has been studied by squeezing flow techniques[85–90]. In such flows (Appendix 11) it is possible to determine the relation between transverse intraply shear strain, shear stress and time scale (Figure 5.10). This figure includes the yield stress data determined from the oscillatory rheometry studies[92], which are found to be in satisfactory agreement with the trends from the large amplitude squeezing studies.

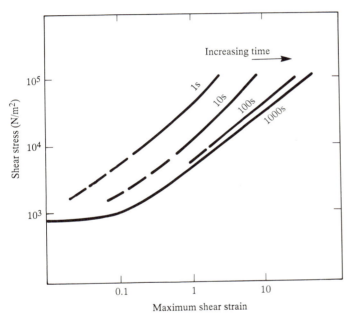

Figure 5.10 Viscoplastic deformation of transverse intraply shear at large deformation, based on the results of Barnes[91]

Axial intraply shear viscosity has also been studied under steady flow conditions[93]. Those data appear consistent with the small amplitude oscillatory studies (Appendix 10).

As well as shear within the ply, it is possible for individual plies to move past one another. This deformation is limited to the resin rich layer in between plies, and is conveniently observed between plies of different orientation. Experimental studies[94–96] suggest that this flow is a true viscous deformation in the resin rich surface layer. Such deformations in practice are often of the order 10 mm. When the ~7 μm thickness of that resin rich interply region is considered, this implies that the shear strain amplitudes in the resin can be of the order 1,000 units of shear. For practical purposes a convenient interpretation is to consider the flow as an interply slip (Figure 5.11).

In this interpretation there is no deformation up to a stress level of about $500\,N/m^2$, but above that shear stress the interply slip velocity increases in proportion to the shear stress. The apparent yield stress may indicate that, at rest, fibres in adjacent plies touch locally, reducing the resin rich layer to nothing. Once stress is applied, these fibres are pushed apart and a uniform, resin rich, lubricating layer is established. Theoretical[97] and finite element[98] studies of this deformation process have also been reported. The latter study indicates how it is possible to apply these data to complex forming processes.

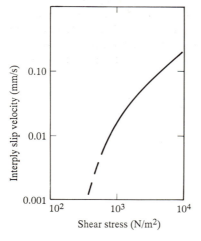

Figure 5.11 Interply slip of CF/PEEK

The optimization of rheology of composite materials requires optimization of the viscosity of the resin and optimization of the distribution of the fibres in the material. Ideally these optimization processes take place at the preimpregnation stage. The consideration of resin viscosity must also take account of the time scales of the process. If the time scale is long and the resin viscosity is low then shaping flows may cause resin to be squeezed out or there may be excessive transverse flow. By contrast, too high a viscosity and too short a time scale means that the reinforcing fibres are unable to adjust to the flow and may become buckled or distorted, particularly on the inside of curved surfaces. Correct tailoring of matrix viscosity and control of resin fibre distribution allow interply slip to occur and the microstructure to be preserved (Figure 5.12). The rheology of the commercial grade of preimpregnated carbon fibre reinforced polyetheretherketone has been tailored to allow it to be shaped in the scales ranging from 0.1 to 1,000 seconds preferably 1 to 100 seconds, while maintaining a uniform microstructure.

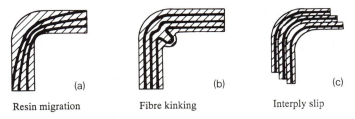

(a)   Resin migration   (b)   Fibre kinking   (c)   Interply slip

Figure 5.12 Influence of viscosity on microstructure after flow: (a) too low, (b) too high, (c) optimized

Besides the constraint imposed by the material, certain geometries of deformation cannot be formed from continuous fibre composites. The theory of the concept of allowable deformations is currently being developed by Rogers[99].

This section has identified and quantified four modes of deformation that can occur with molten thermoplastic composites: resin percolation, transverse intraply shear, axial intraply shear and interply slip. Together with the concept of allowable deformations[100], these determine a hierarchy of deformation processes of increasing complexity (Figure 5.13).

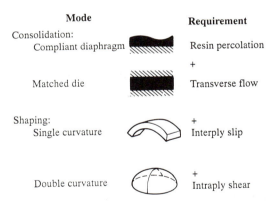

Figure 5.13 A hierarchy of deformation processes

5.1.4 The analysis of processing operations

Having established the basic physical properties of the material that are relevant to processing operations, we must consider how to model those technologies. An excellent modelling framework has already been established[101–105]. By empirical studies of the various forming processes with constant reference to the basic physical properties of the material, which are independent of the technology, it should be possible to refine those models to achieve sound predictive capacity. In respect of heat transfer[106–112] such divination is already possible. For shaping flows, and even for consolidation, we still have a long way to go before a full consensus is achieved, but the auguries must be considered fair. The plastics industry has already established predictive capability for highly complex processes such as injection moulding, and, although different, the processes of composite manufacture should be amenable to a similar analytical approach.

5.2 Manufacturing technology

The processing stage is the definitive operation that distinguishes between a thermoplastic composite and a thermosetting one. In the case of thermoplastic materials, manufacturing of the end product can be completed in a matter of seconds. The processing stage requires only heat and work. It is an entirely physical operation: all the chemistry has been carried out by those people whose business is chemistry.

When first introduced[113], continuous fibre reinforced thermoplastic materials, such as carbon fibre reinforced PEEK, were noted to be stiff and boardy prepregs, lacking the tack and drape which were considered indispensable to the hand lay up technologies established for thermosetting composites. The unfriendliness to hand lay up is, however, offset by a strong suitability to automated processing, and especially the adoption of variants of metal working technology[114]. The absence of tack also means that thermoplastic composite materials do not require the stringently clean room conditions that are necessary for the manufacture of high quality thermoset laminates. In addition, the assured quality of fully polymerized materials allows the processor to remove some of the peripheral equipment, such as refrigerated storage, which is an indispensable adjunct to thermosetting processing[115].

With thermoplastic composites there are no considerations of a 'lifetime' in which the prepreg must be moulded. The difficulties associated with a lack of tack and drape can be overcome. It has been demonstrated that thermoplastic composite materials can be processed by the same hand lay up and autoclave technology favoured for thermosetting systems. Certain thermoplastic product forms, such as commingled fibres, possess drape, and tackifying agents can be added to powder prepregs. Such alternative product forms are discussed in Chapter 3. When using hand lay up technology, the processing costs for thermoplastics are similar to those for thermosets, shorter processing times being offset by difficulties of higher processing temperatures. When using hand lay up, the main advantages in using thermoplastic materials come from their service performance. The economic advantages for thermoplastic composites are derived by developing the high rate, automated, technologies that exploit the inherent rapid processability of the material.

A number of general reviews[116–125] consider the range of processing operations available with this class of materials. Those technologies include:

consolidation in matched die or autoclave,
tape placement by filament winding or tape-laying,
continuous forming by pultrusion, roll-forming or double belt processing,

stamping, especially adaptations of metal working,
sacrificial diaphragm-forming,
sequential processing operations,
assembly by conventional fasteners adhesives or designed thermoplastic technology,
machining,
rework, repair and reclaim.

Several of these papers emphasize that it is desirable to integrate innovative design with innovative processing in order to reduce the part count in a structure to a minimum and achieve thereby the most cost effective structures. Such little data on economic aspects as has been published in the open literature portrays a consistent picture. Duthie[126] and Archibald[127], working from quite different standpoints, conclude that, with thermoplastic processing, it is possible to achieve structures that not only save weight in comparison to metal but involve only half the cost. Foley and Bernardon[128] emphasize that low cost benefits with thermoplastics are achieved through the application of automation.

5.2.1 Consolidation

The simplest process is the consolidation of laminates. This is an integral part of all forming processes and, since flat sheet is also an important product form of these materials, it is a process worthy of study in its own right. Because the properties of the moulding depend upon the quality of lamination, it is essential that we define the conditions under which well consolidated material can be obtained.

Lamination comprises two processes: obtaining auto-adhesion between the laminates, and removing what lies between them. It is, of course, obvious that it is not possible to start to fuse the layers together until they are immediately juxtaposed physically. The desirability of removing air from that interface is often ignored. Thus Lee[129], who designed an experiment to determine the auto-adhesion of carbon fibre/PEEK laminates, separated two plies by a thin foil and measured the build up in strength as a function of time. He concluded that it took approximately 2 hours to achieve full consolidation, and interpreted this as being the time scale for auto-adhesion. An alternative interpretation of this experiment is that he was measuring the time scale for the air trapped in the blister to be dissolved into the molten polymer. Subsequent experiments by the same group[130] have demonstrated auto-adhesion of the same material in less than 5 seconds. In the case of preimpregnated materials there is only the air between the layers of prepreg to be considered. An altogether larger volume of air more intimately associated with the materials is found with the various drapable product forms, and failure to take proper precautions to remove that air may be the cause of some of the negative comments about the quality of laminates moulded from such materials[131-3]. There are two methods by which air can be removed: pressure to dissolve it into the melt, and vacuum to extract it in the first place. These methods must be supplemented by consideration of space and time. In the moulding of large area laminates it is desirable to design the process so that equilibrium conditions are approached by a pressure and temperature gradient across the moulding, with the temperature and pressure rising to the processing conditions in the centre of a large moulding before the edge. These desirable features, in particular the temperature gradient, are often natural features of a process. Cattanach[134], who was responsible for much of the early work in developing processing technology for this class of material, particularly emphasizes that, in any consolidation process, the first and most vital question to ask is how the air is to be displaced.

Before considering the auto-adhesion process it is necessary to consider if any flow is required before consolidation can occur. Ideally, if the prepreg surfaces were mirror flat and perfectly parallel, we could proceed immediately to auto-adhesion. This ideal case is not usually met with, and when two plies, particularly plies of different orientation, are brought into contact the roughness of those surfaces will leave spaces to be filled. These spaces can be filled by resin squeezed out of the prepreg[135] and, in addition, some transverse rearrangement of the fibres may be required[136]. Such transverse flow becomes essential if the consolidation is between matched dies[137].

Many authors[138–156] have made systematic study of the consolidation processes of preimpregnated carbon fibre/PEEK in presses and autoclaves. Recommended moulding conditions are indicated in Appendix 12. In respect of temperature most authors demonstrate a preference for the temperature range 360°C–400°C, as suggested in the manufacturer's data sheet[157]. While confirming that preference, Kempe and Kraus[158] demonstrate satisfactory moulding at 420°C. At the other end of the spectrum Nagumo and his colleagues[159] indicate satisfactory mouldings after preheating to 350°C and pressing at 310°C. A consensus view would suggest that 10 atmospheres ($1 MN/m^2$) is a preferred pressure for consolidation: Iaconis[160] shows no effect of pressure over his range of study from 3 to 40 atmospheres; Manson and Seferis[161] confirm this over the range 2 to 10 atmospheres; and Kempe and Krauss[162] indicate satisfactory consolidation under vacuum only (1 atmosphere).

A difficulty associated with high consolidation pressures is the tendency for transverse flow to occur[163]. This is an especial problem if several plies of the same orientation are stacked together. In such cases lateral constraint must be provided. High pressure can also cause resin to bleed out at the free ends of the moulding[164]. This demonstrates the axial permeability of the fibre bed (Section 5.1). For preimpregnated materials there is no gain from the use of consolidation pressures above 10 atmospheres: this is not the case with the various assembled product forms, where such pressures are generally regarded as sub-minimal. Times for full consolidation are variously reported. Most authors indicate a preference for time scales of the order 5 to 30 minutes, but this preference is almost certainly a matter of experimental convenience, being the time taken to allow thermal equilibrium in the moulding. An early report by Kempe[165] indicated full consolidation of molten prepreg in less than 1 minute, while the latest results from the Stanford team indicate less than 5 seconds[166]. At the long time scale Manson[167] demonstrates no ill effects after 2 hours, while Kempe and Krauss[168] have extended their study to 3 hours without unacceptable degradation. In respect of time, temperature and pressure – as with the effects on morphology associated with different cooling rates after moulding (Chapter 4) – the manufacturer's recommended moulding conditions[169] appear conservative in respect of the breadth of processing window for fully impregnated carbon fibre/PEEK.

The insensitivity of the consolidation process for preimpregnated carbon fibre/PEEK should not be source for surprise. As indicated in Section 5.1.3, the relaxation time of molten PEEK as used in composite materials is of the order 0.1 seconds and is not significantly effected by temperature or pressure in the range of interest. This time scale represents the range where deformation changes from dominantly elastic to essentially viscous flow. At substantially shorter time scales, say 0.03 seconds, the molecules will not be able to relax stresses by self diffusion, while at longer times, say 0.3 seconds, essentially viscous flow will have occurred with full self diffusion. Provided full contact is made between the surfaces, auto-adhesion should occur in less than half a second.

One further factor of importance in the manufacture of simply consolidated parts is the tooling. It is desirable that the tooling should have a similar thermal expansion to the material being moulded, especially where shaped mouldings are concerned. Tooling materials have been specially developed to meet this requirement at the high processing temperatures of thermoplastic composites[170–171]. In particular, integrally heated tooling of low thermal expansion has been demonstrated to be of value, in conjunction with a simple press clave[172]. The use of such a combination is of advantage when using vacuum assisted processes, because of the improved freedom which it gives for securing the bagging materials in an unheated part of the mould. An alternative approach to using materials of low thermal expansion is to reduce the temperature excursion of the moulding tool. This can be achieved by transferring a preheated blank into a 'cold' mould, and then ejecting. Such a transfer process is the basis of many of the adapted metal working technologies[173–174], and also leads to a rapid cycle time in simple sheet forming. Tooling, and accessories such as bagging materials (Appendix 13) are important elements in the design of optimized processing.

A typical consolidated laminate is shown in Figure 5.14.

Figure 5.14 Laminated sheet stock in CF/PEEK

Matched die or autoclave moulding of simple shapes such as 'V' sections or half cylinders are obvious extensions of simple lamination.

Hollow sections can also be formed by consolidating on to an internal mandrel. Here the low thermal expansion of the composite material is seen to advantage, since a metal mandrel will then shrink away from the moulding and may easily be extracted. Hollow mouldings can also be moulded around soluble cores that can subsequently be dissolved out of the moulding. Such a soluble core material can be selected to have a high thermal expansion coefficient and therefore aid in maintaining a uniform internal pressure on the composite as it is formed. While it is usual to consolidate on to a tool, it is also possible to consolidate a composite inside a closed mould by inflating a bladder (Figure 5.15). This technology is advantageous when a precisely defined external section is required.

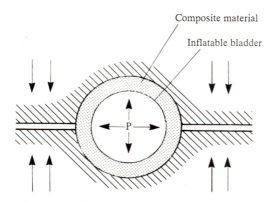

Figure 5.15 Bladder forming

Lamination is the simplest, and the most critical, processing operation. Whatever product form is to be fabricated, mastering the art of simple consolidation is an essential step on the learning curve.

5.2.2 Continuous consolidation

Continuous consolidation of simple laminates can be carried out by a variety of technologies. A natural extension to simple batch pressing is to operate a step pressing process, by which a preassembled sheet stock is fed through a preheating oven or press and a cold consolidating press (Figure 5.16). The true continuous variant of this intermittent process is double belt lamination (Figure 5.17).

This process has been demonstrated by Ives[175] and Kenworthy and Cattanach[176], and has also been the subject of theoretical study by workers at Georgia Institute of Technology[177,178]. Speed in such processes depends upon the length of the band and the rate at which thermal equilibrium can be achieved, noting that both a heat

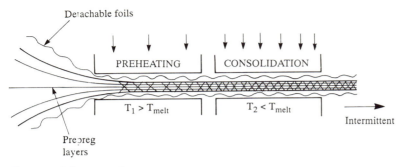

Figure 5.16 Step pressing

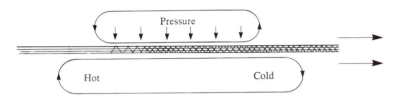

Figure 5.17 Double belt lamination

up and cool down section are required. While such processes are naturally suited to the production of flat sheets, they can be adapted to produce curved laminates by a designed imbalance of the lay up (Chapter 4). The main constraint at the present time (1990) on the development of continuous sheet consolidation techniques is the automatic lay up of the preimpregnated tape. Various theoretical[179] and practical solutions to this issue have been discussed. As with any automated technology, a heavy investment in capital equipment is required in order to produce a high volume of output at low cost. The technology is poised ready to meet the demands of mass production.

Rod stock and simple shapes can be made in continuous pultrusion processes[180–184]. The processes can use either preimpregnated tape or work direct from the resin and fibre feedstock. Commercially available products are of small cross sectional area, typically less than 10 mm square, but the technology is capable of much larger scale. Thicker sections inevitably require more powerful haul off equipment, and so higher capital investment. A particular advantage of pultruded thermoplastic products is that they can be post formed in subsequent processing operations. Pultrusion has also attracted particular interest as a process by which large space structures can be fabricated in orbit[185,186]. A common use for thermoplastic pultruded product is as a fillet in T-shaped stiffeners (Figure 5.18).

Pultrusion is most simply carried out with uniaxial product. Woven fabrics, and more particularly braided[187] product forms, can be advantageously combined with uniaxial material to achieve torsional as well as axial stiffness in a pultruded component.

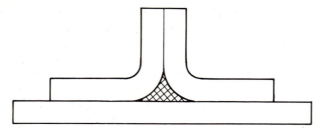

Figure 5.18 T stiffened panel with pultruded fillet

5.2.3 Tape placement

Tape placement, as expressed by tape laying of flat, or nearly flat, sheets, and filament winding is the most natural expression of continuous fibre technology. This technology contains the seeds of automation, and the ability to define and translate fibre orientation from the design concept to the finished structure. Well established for thermosetting materials, it can be adapted to thermoplastics by the addition of a heated applicator head.

Tape laying for thermoplastic prepregs was first described by Brewster and Cattanach[188]. The principle of their design is shown in Figure 5.19.

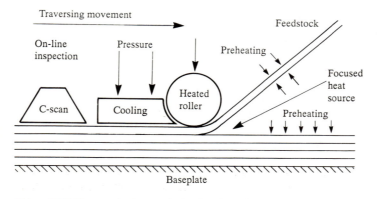

Figure 5.19 Thermoplastic tape laying head

Much of the art of thermoplastic tape laying is in the heat exchange process. This aspect has been studied by Grove[189] and by Beyeler[190,191] who recommended laser heating of the tape. Each layer of prepreg tape is melted and consolidated in place. At each application part of the ply underneath is also remelted. As the layers build up, the lower layers of the moulding cease to be heated: there is a tendency to end up with a moulding where the final cooling process is from a temperature gradient through the moulding. This uneven cooling may cause internal stress in the moulding and lead to distortion. Since the majority of those thermal stresses arise between Tg and ambient (Chapter 4), the use of a baseplate heated to above Tg can reduce this problem. Anderson and Colton[192] emphasize the desirability of preheating the surfaces to be fused. They find improvements in consolidation at pressures up to 10 atmospheres. Grove and Short[193] stress the desirability of maintaining tension on the tape during processes, so as to avoid kinking of the fibres (Chapter 2). Both Anderson and Colton[194] and Grove and Short[195], report approximately 75% consolidation, and this is consistent with industrial experience. The lack of complete consolidation and the problems of thermal distortion mean that, at the present time, this technology is used as an automated lay down strategy for subsequent full consolidation in another process. Indeed partial consolidation is seen as a positive benefit in the feedstock for processes such as diaphragm forming (section 5.2.6), because it allows the additional freedom of interply slip before final consolidation. The tape laying process is seen as a starting point to preconsolidate laminates for subsequent stamping[196].

The associated technology of filament winding, and its particular significance to thermoplastic technology, is considered by Bowen[197] and Mahlke[198]. In comparison with tape laying, filament winding has an advantage in that tape tension can be geometrically translated into consolidation force. This is most particularly true with small diameter tubes. The disadvantage of filament winding in comparison to tape laying is that the traversing of the tape necessarily produces a cross-over and potential void (Figure 5.20).

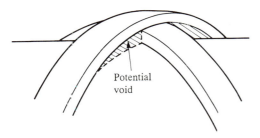

Potential
void

Figure 5.20 Cross-over void in filament winding

This defect must be avoided by local transverse flow or resin permeation. The forces to achieve such flows are preferably introduced by a consolidation roller. The void content in typical parts is reported to be in the range 1.5 to 5%[199–202]. This can be reduced to less than 1% by subsequent consolidation in an autoclave[203]. A range of heating technologies has been used. These include laser[204], hot gas, hot shoe and infra red. Each approach has its advocates. Rowan and Askander[205] and Dickman[206] suggest suitable winding speeds up to about 0.5 m/s to obtain good consolidation: above this threshold speed the quality of *in-situ* consolidated components deteriorates. This figure is of interest in the consideration of consolidation time. If we assume that the tape remains molten for about 50 mm length after making contact with the substrate, the time interval available for auto-adhesion at a winding speed of 0.5 m/s is of the order 0.1 second, which compares satisfactorily with the estimate in Section 5.2.1 above. The theoretical modelling of the filament winding process has been undertaken by workers at the University of Delaware[207] and the heat transfer process has been particularly studied[208]. Filament winding is now established as a manufacturing process for high quality components (Figure 5.21).

Figure 5.21 Filament wound structure by ICI Composite Structures

In comparison with thermosetting filament winding, four advantages are apparent for thermoplastic technology. First is the potential for non-geodesic winding patterns to be established[209]. The second advantage is the ability to post form the structure, including giving it a re-entrant[210]. Filament wound sections can be remoulded in other ways, for example, a box section can be cut in half and reconsolidated to form an 'I' beam. The final advantage of thermoplastics is that there is no need for an autoclave. This feature, which compensates for the relatively slow winding speed, is particularly exploitable in the manufacture of very large structures that would otherwise require impossibly large autoclaves. An additional feature of large thick structures is that it may not be convenient to wind them all at one time: with thermoplastic materials an interruption in the process has no effect – there are no considerations of shelf life.

5.2.4 Continuous forming

Besides continuous processes such as double belt lamination and pultrusion (Section 5.2.2 above), it is possible to adopt certain well established, metal working, continuous forming processes to work with thermoplastics. Of this family, roll forming[211,212] (Figures 5.22 and 5.23), has received particular attention.

In this process a preconsolidated sheet stock is preheated to above the melting point and fed into a series of cold roll sections. In this series of, typically five, matching rolls the shape is progressively formed and then stabilized. Ideally the blank is shielded, so that only those sections of the moulding which are to be

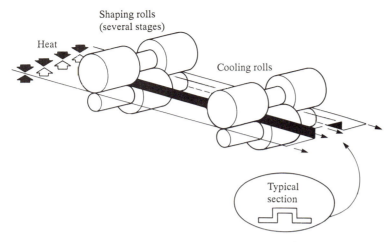

Heat

Shaping rolls
(several stages)

Cooling rolls

Typical
section

Figure 5.22 Schematic of roll-forming process

Figure 5.23 Roll forming line[213]

shaped are heated. The rollers more readily grip the solid, unheated parts of the sheet. This process is particularly suited to the formation of simple channel, or hat, sections (Figure 5.24).

Line speeds of the order 10 m/min have been reported for this process. This process is suitable for straight channel or hat sections, but Cattanach[214] has also demonstrated that it can be used to manufacture curved beam sections (Figure 5.25).

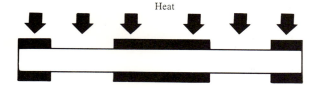

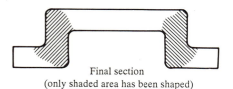

Final section
(only shaded area has been shaped)

Figure 5.24 Formation of a 'top hat' section

Figure 5.25 Roll formed sections

5.2.5 Stamping

Another technology adopted from metal working is stamping of a preheated blank in a cold tool[215–220]. Various options for tooling have been adopted, including matched dies, rubber blocks and hydro-rubber forming (Figure 2.26).

Prototype mouldings have also been successfully formed into wooden tools. At first sight it is surprising that it is possible to mould a material at 400°C in a rubber or even wooden mould. However, the low total heat content of the thin sections of

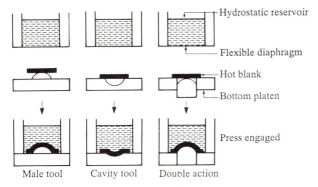

Figure 5.26 Schematic of hydro-rubber forming

composite material, combined with relatively slow heat transfer into a mould of high thermal mass, ensures that even rubber blocks or diaphragms have a useful life. This technology is best suited to the forming of simply folded shapes requiring only a minimum of deformation in the material: since the time scale of forming, typically 1 second or less, is so short, there is little time for axial or transverse intraply shear to occur, so that the primary mode of deformation is constrained to interply slip (Figure 5.27). Because deformation is limited in this way, there is good definition of part thickness.

The main art of such rapid stamping for complex mouldings is cutting the blank from which the component is stamped. This art is also shared with the metal-working skills from which the technology has been borrowed.

Figure 5.27 Hydro-rubber formed rib section

For the rapid stamping of deep drawn components, some workers prefer to use a woven fabric form. The fabric constrains the deformation and also stops uneven thinning of the material by preventing any significant transverse intraply shearing and, in particular, any splitting of the preimpregnated tape.

The technique of hydro-rubber forming is a high pressure process but is limited in scale to about 0.25 m diameter section. Much larger scale components can be moulded with rubber block processing. Workers at Westlands[221] have demonstrated mouldings up to 3 m long.

Stamping preheated, preconsolidated sheet can give precision shaped, simply formed, components of excellent quality at cycle times of less than a minute. As such, it is a preferred process for the transition from bespoke hand lay up structures to mass production technology.

5.2.6 Diaphragm forming

Certain processes for thermoplastic composite fabrication, such as autoclave consolidation, filament winding and pultrusion, have been adapted direct from thermoset technology; and others, notably roll forming and rubber block stamping, owe their genesis to the technology of metal working. Optimized thermoplastics processing technology will be developed specifically to work with this new class of materials.

One such technology, designed to mould large area structures with pronounced double curvature, is diaphragm forming. This technology, originally described by Barnes and Cattanach[222], has been the subject of extensive study by researchers at the Universities of Delaware and Galway[223-236]. A development of that technology, to include rapid forming of preheated material on to cold moulds – the 'Diaform' process – has been reported. The range of components necessary to assemble a fighter's forward fuselage have been formed by this technology[237].

The plan of diaphragm forming (Figure 5.28), is to sandwich the prepreg layers, freely floating, between two constrained diaphragms.

The air between the diaphragms is evacuated and the laminate is melted. Pressure is then applied to one side to deform the diaphragms and cause them to take up the shape of the mould. Because the prepreg layers are freely floating, and, when molten, are less stiff than the diaphragms, they conform to this moulding process.

Once forming is complete, the moulding is cooled. The diaphragms are then stripped off, and the component can be trimmed to size. This technology has proved especially valuable in forming large area double curvature structures – (Figure 5.29).

The process was originally developed to utilize superplastic aluminium diaphragms[238], but latterly most emphasis has been placed on the use of polymeric diaphragms, for which 'Upilex' is the preferred material[239]. It is the diaphragms which control the forming process, and the stiffness of the diaphragm is a critical factor in determining the optimum process. For simple component shapes, compliant diaphragms are to be preferred, but, if the shape becomes complex,

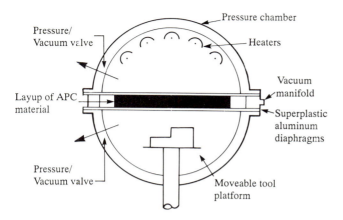

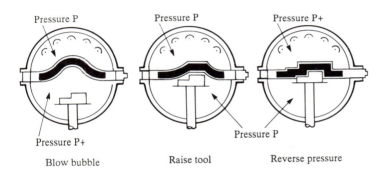

Figure 5.28 Principle of diaphragm forming

Figure 5.29 Aircraft fairing made by diaphragm forming

these may allow buckling in the moulding. The more complex the moulding, the stiffer should be the diaphragm, and so the higher is the required moulding pressure. High moulding pressures can induce significant transverse squeezing flows in the composite, and so produce variations in thickness through the moulding (Figure 5.30).

Figure 5.30 Excessive transverse flow when using very stiff diaphragm

Other important variables in the process include the forming rate[240] and the area of the final moulding in comparison with that of the sheet from which it is formed. O'Bradaigh[241] points out that slower forming rates, and a close match between the areas of the blank and the formed shape, are to be preferred, in order to avoid out of plane buckling. Precise sizing of the blank is also a desirable characteristic economically. Forming rate is seldom a constraint unless cold tooling is to be used. The diaphragm forming process can be used to make full hemispheres from flat sheet with no significant distortion of the reinforcing fibres. Moulding cycles last between 10 minutes and 2 hours; dependent on the tooling, heating and cooling methods to be employed, the actual time to shape the moulding is typically 1 to 5 minutes.

Many studies of diaphragm forming have utilized autoclaves, because they are the standard instrument for thermoset technology. Much faster processes are achieved by the use of a separate heat source and pressure vessel. The use of heated tooling is also a desirable feature, as is the ability to cool the tool rapidly. Mallon and O'Bradaigh[242] have successfully used lightweight tools made from sheet steel, which facilitate rapid cooling.

In the diaphragm forming process the floating composite material is free to shape itself in all the allowable modes of deformation. This freedom necessarily results in significant reorientation of the reinforcing fibres[243,244] and models to predict that reorientation are being developed. The use of tracer fibres and X-ray analysis shows that that reorientation, if not yet completely predictable, is reproducible.

In the early stages of the life of a new material it is generally desirable to exploit adaptations of established processing technologies in order to gain acceptance of those materials, but the full exploitation of new materials in service usually requires the development of bespoke processing methods if those systems are to develop their full economic potential. Diaphragm forming is the first technology specifically designed to process the new generation of thermoplastic composite materials. The versatility that thermoplastic composite materials have demonstrated in manufacturing suggests that there is opportunity creatively to exploit the forming properties of thermoplastics even more than the industry has done to date.

5.2.7 Incremental processing

Linear chain thermoplastic polymers may be reprocessed a number of times. For example, a flat sheet structure can be locally heated along a line and bent to form an angled structure, or a filament wound tube can be locally reheated and the convex curvature remoulded to form a concave section[245] whose direct winding would prove very difficult. Strong and Hauwiller[246] demonstrate how this principle of incremental forming can be exploited in the moulding of large structures, using small, low capital cost, fabrication equipment.

Incremental processing of thermoplastics can also be used to increase the versatility of stock shapes. Linear structures, such as pultruded sections, can be reheated and formed into curved sections or their ends can be reshaped to facilitate joining to other structures. The combination of incremental forming with the economics of scale of stock shapes allows the possibility of manufacturing bespoke tailored structures of aesthetic design from high performance composite materials at reasonable cost. This inherently thermoplastic design philosophy has yet to take root in the industry.

The concept of incremental forming with thermoplastics can be extended to product form. First is the integration of two thermoplastic composites, based on different fibres, into the same moulding. As an example of this, a tubular structure requiring axial stiffness and transverse toughness can be formed from alternate layers of axially laid, ultra high modulus composite prepreg, overwrapped with hoop wound, high strength fibres in the same matrix. Composite material can also be incorporated on to a neat resin or short fibre moulding, in particular a tubular structure, as an overwrap. A similar approach to achieve selective reinforcement is to incorporate continuous fibre composite material as an insert in a conventional moulding of resin or short fibre compound as selective reinforcement. Generally such reinforcement can be incorporated in an injection moulding or stamping process. These latter approaches allow the designer to uprate the performance of conventional mass produced plastic mouldings. Where this approach is to be utilized, the main considerations are to obtain adequate heat transfer to achieve true melting across the interface of the product forms, and to balance the moulding to accommodate the differential thermal expansion of the different product forms.

Thermoplastic composites can also offer the potential to integrate the production of sandwich structures with foam formation[247].

5.2.8 Machining

The toughness of thermoplastic composites permits the ultimate liberty of machine finishing a moulding. The first requirement for successful machining of thermoplastics is to keep the workpiece cool. Water jet cutting is successful with thermoplastic composite materials in contrast to laser cutting, which tends to melt and char the workpiece. Conventional metal-working tools can be used successfully, provided plenty of water is used, but there is a strong preference for diamond tools to avoid the problem of excessive tool wear.

Machining, especially trimming the edges of a moulding, and cutting out necessary holes and inserts, is a standard part of composite processing. With conventional low toughness composites, such operations can cause delamination of the workpiece and must be carried out with great care. The toughness of thermoplastic composites allows a more vigorous approach, consonant with the needs of mass production technology. That toughness also allows intricate machining of structures, including the cutting of threads[248] (Figure 5.31).

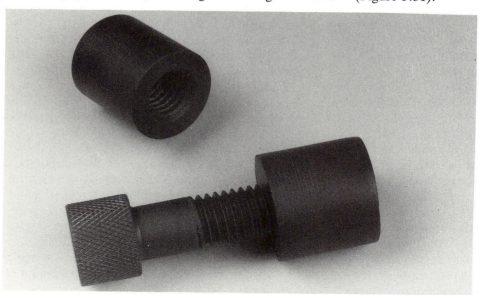

Figure 5.31 Nut and bolt machined from carbon fibre/PEEK

A dramatic example of the machinability of thermoplastic composite materials, using conventional metal working technology, is the ability to punch out small components from thin sheet at room temperature (Figure 5.32).

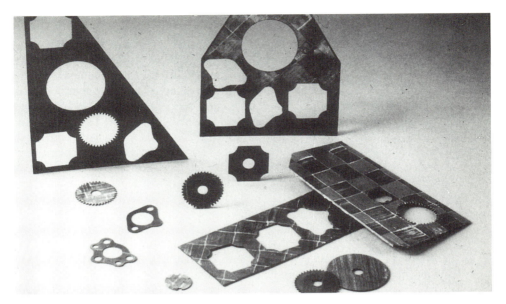

Figure 5.32 Components punched from sheet

A corollary of easy machining with thermoplastic composites is the ability to incorporate inserts into the structure where these are required for a specific purpose. Usually such inserts are metallic and, in this case, it is necessary to take into account probable differences in thermal expansion of the materials, especially if the component is to see service at other than ambient temperature. Usually, in comparison to the metallic insert, the thermoplastic composite material will be of lower thermal expansion in the plane but higher thermal expansion through the thickness. In the case of above ambient service temperature, this differential expansion will tend to increase the tightness of the insert.

Machining, and the particular ways in which this most versatile of all manufacturing technologies can best be exploited with thermoplastic composites, is deserving of a more detailed technical study than they have so far received.

5.3 Assembly technologies

There are two classes of moulding. The first represents a finished structure and is complete in itself. The second is only a component of a structure and requires assembly to perform its function. It is desirable that the inherent properties of the structure are not compromised by this assembly process. Further, this operation must be accomplished with a minimum penalty in respect of cost and weight.

With thermoplastic composites, there are four assembly strategies: fasteners, adhesives, fusion bonding and interlayer bonding[249]. Each approach has its place depending on the nature of the assembly and its function in service.

5.3.1 Fasteners

Undoubtably the most widely used assembly technique for advanced structural composites is the use of fasteners[250]. This is particularly indicated where the component is required to be able to be disassembled in order to gain access to the interior of the structure. When the composite material is to be assembled to a dissimilar material, for example a metal substructure, fasteners, in the form of rivets, are also widely used. Because fasteners will usually include a small amount of compliance, their use allows conformation to differences in thermal expansion and distortion under load of the different components of a structure. Rivets are also often included as a secondary assembly technology with bonded structures; they supply an element of through thickness reinforcement, which resists out of plane loading that might otherwise cause delamination. Walsh[251] provides a comparative assessment of bolted joints in thermoset versus thermoplastic composites, and concludes that, despite using a higher performance fibre with the epoxy matrix, the PEEK based composites show improved bolted joint performance.

There are several problems associated with the use of fasteners. Because they nearly always require a hole to be made in the structure, they necessarily weaken it. For structural application a very large number of fasteners, usually every 50 mm, are required. On large structures there may be more than one line of such fasteners. The cost of inserting such large numbers of fasteners is a significant element of the total cost of a structure in service. Generally the fasteners are metallic, and so add to the weight of the structure. Despite these problems, so great is the experience of the use of fasteners that they will undoubtably continue to be used extensively.

Thermoplastic composites themselves can be used as fasteners. Pultruded rod stock provides a simple rivet whose end can readily be fused into a cap (Figure 5.33).

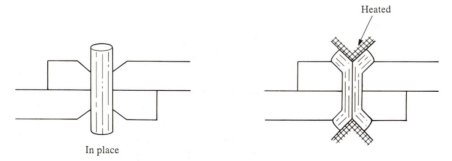

Figure 5.33 Thermoplastic composite rivet

The thermoplastic rivet can also be directly heated, for example by resistance heating (see Section 5.3.3), in order to integrate it with the components. The use of a composite fastener eliminates the weight penalty associated with conventional systems.

5.3.2 Adhesive bonding

Since the epoxy resins used as the matrices for thermosetting composites are also widely used as adhesives, it is not surprising that adhesive bonding plays a major role in composite technology. The classical adhesive bonding technology for thermosets can also be applied to thermoplastic composites[252–262]. However, the chemical inertness of thermoplastic composites means that special attention must be paid to surface preparation in order to obtain optimum bonding.

The search for optimum surface treatment has produced a diversity of different answers:

   sodium hydroxide etch[263,264]
   grit blast and wipe clean[265]
   strong acid etch[266,267]
   plasma treatment[268]
   silane coupling agents[269]
   corona discharge[270,271]
   'Kevlar' peel ply[272]

These methods range from purely physical to chemical attack, where the objective is to achieve a strongly activated surface. The diversity of approaches suggests that there is still room for optimization, and that there is plenty of freedom of manoeuvre. From reading these papers one gets a general impression that, with appropriate surface treatment, adhesive bonding with conventional film adhesives will give as good a bond with thermoplastics as with thermosetting composite systems.

Naturally the use of epoxy adhesives introduces into the structure some of those features – susceptibility to moisture and damage – that thermoplastic materials seek to overcome. Many thermoplastic resins, and PEEK in particular[273], are themselves excellent hot melt adhesives.

5.3.3 Solvent bonding

A conventional method of assembling components made from linear chain thermoplastics is to dissolve the surface layers to be joined with a solvent, and then fuse them together. The residual solvent diffuses into the rest of the moulding, but is not of sufficiently high concentration significantly to soften that material. Ultimately the residual solvent evaporates from the structure. To aid the fitting together of the two surfaces, which will have minor irregularities, the medium to dissolve the surface is usually a solution of the polymer. This technique is widely known to the modelmaking fraternity in the assembly of polystyrene kits.

In the field of structural composites this approach can be used with amorphous matrix materials where convenient solvents exist, although at this time I know of no publications describing an optimization of this technique. It can, in principle, also be used with semi-crystalline polymers, where the solvent must usually be a strong acid. PEEK composites have been bonded by this method, using concentrated sulphuric acid[274], but no champions of the technology have yet appeared.

5.3.4 Fusion bonding

With thermoplastic materials, there is the potential for direct fusion bonding of components. Co-consolidation of shaped components, by remelting the whole structure, requires expensive tooling to support the structure while molten. This can be avoided by locally heating the surfaces to be joined. Usually an additional layer of neat resin is included to promote 'fit up' between the components. Optimum fusion bonding techniques develop a weld strength equivalent to that of the fully consolidated composite, so that the mechanical performance is determined by the composite properties and joint geometry only.

To achieve local fusion of the bond line, a wide variety of techniques have been explored. The literature contains several comparative reviews[275-282] and detailed studies[283-293]. The options include:

resistance heating[294-300]
induction heating[301-306]
microwave[307]
infra-red[308]
ultrasonics and vibration[309-314]
platen heater[315]
heated inserts[316]
hot gas[317]

From this richness of choice, resistance heating, ultrasonics and induction heating stand out as preferred technologies.

Ultrasonic vibration has several champions. Commercially available equipment can be used, but the size is limited, so that this technique is really only suitable for small mouldings or the moulding must be fused in incremental steps. To obtain the most satisfactory results, the faces to be joined should be grooved. This allows the energy to be initially focused into a series of point contacts.

As with other fusion bonding techniques, the quality of the bond is enhanced if the surfaces to be joined are of neat resin. Component preparation should preferably include an integrated surface layer of pure polymer. In the case of carbon fibre/PEEK composites that surface layer can be made from commercially available 'Stabar' PEEK film.

Induction heating also utilizes a well established technology. Heating can be directly induced in carbon fibre composites because of the conductivity of the fibres. The strongest heating effects are induced in cross plied laminates or woven fabrics. It is difficult to induce significant heat generation in uniaxial laminates. The most satisfactory method is to incorporate a ply of metallic wire mesh embedded in neat resin to focus the induction heat in the critical region.

The most widely explored of the direct fusion bonding technologies utilizes the electrically conductive properties of carbon fibre preimpregnated thermoplastic tapes to obtain a resistance heating element. This process is shown schematically in Figure 5.34.

There are several important details in this process. The electrical connection to the prepreg tape is a critical region. That connection is impeded by the insulating

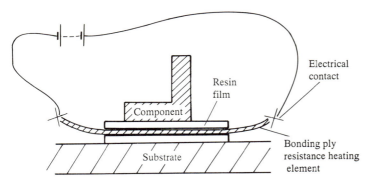

Figure 5.34 Schematic diagram of resistance heating using carbon fibre prepreg. Note that typical component and substrate thicknesses are of the order of 2 mm, and the resistance heating element and resin films are of the order of 0.1 mm thick

resin in the prepreg, and ideally this should be removed. One method of obtaining a good connection is to pyrolize the resin at the end of the strip and press softened solder into the exposed fibres[318]. Thin layers of neat polymer film on either side of the resistance heater have an important function in insulating the workpieces from the resistance heater. If these insulating layers are omitted, then the current can pass direct into the workpiece and become dissipated. As a result, any fusion is limited to the ends of the workpieces. Booth[319] has demonstrated that cooling of the workpieces also assists in achieving longer and more uniformly strong bonds. From the extensive studies in the open literature it is now possible to design resistance heated bonds with confidence. The technology can be automated[320], and is particularly suited to the attachment of stringers to plates. For bonding large areas, it is desirable to use an incremental method. By bonding only a small area at a time the process can more readily be controlled, and a modest power supply can be employed. It is also a technology that can make use of readily available heating sources, and therefore has useful potential for field repair: the first demonstration of resistance welding of carbon fibre/PEEK was powered by a 6-volt car battery[321].

Direct fusion bonding techniques are capable of developing the full strength of the composite material in the weld. With thermoplastic matrices, this includes the advantages of environmental resistance and of damage tolerance.

5.3.5 Interlayer bonding

A novel thermoplastic technology, specifically designed for use with high-performance thermoplastic composites, incorporates a layer of lower melting point material on to the surface of the composite to be bonded. The workpieces can then be fusion bonded below the melting point of the structural components. These remain form stable throughout the process: no complicated jigging is required[322–324]. Ideally the low melting point interlayer should have a glass transition temperature equal to or greater than that of the resin in the composite material. The interlayer resin should preferably be miscible with that of the composite. During the preparation of the component incorporating the interlayer, it is desirable that there should be some migration of the reinforcing fibres in the structural composite into the interlayer region. To achieve this latter requirement, it is essential that the interlayer resin is fused to the composite component at a temperature above the melting point of the composite matrix resin, ideally during the consolidation of the sub-components. Meakin[325] demonstrates how these desirable objectives can be simultaneously achieved with carbon fibre/PEEK composites in conjunction with a polyetherimide film (Figure 5.35) – the 'Thermabond' process.

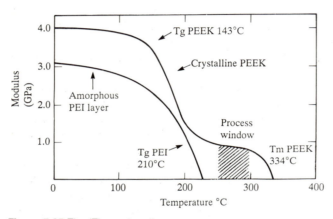

Figure 5.35 The 'Thermabond' process

The broad temperature range over which the interlayers can be fused provides a robust process[326,327] where the full properties of the composite can be achieved. Further, it is possible to use the local fusion bonding techniques described in the previous section where they are appropriate[328,329], although direct heating is usually preferred. An additional layer of interlayer film may conveniently be incorporated into the joint to aid 'fit up' of the components. Silverman and Criese[330] have compared 'Thermabond' with other bonding technologies for the criteria of reproducibility, adaptability to various joint configurations, suitablility for small or large area bonding, minimal surface preparation, minimal equipment use, adaptability to automation, on-line inspection capability, reworkability and impact on the production environment. 'Thermabond' was the only technology to which they awarded 100%.

One potential concern with the 'Thermabond' technology, sometimes described as amorphous interlayer bonding, is the introduction of a thin layer of amorphous material that may be susceptible to solvent attack. Smiley and Halbritter[331] have demonstrated that there is no degradation of properties of the bondline due to water, and that useful strengths are retained on exposure to a range of solvents. In addition, because the bondline is thin, it is possible to consider protecting such areas from attack. Meakin[332] has also demonstrated that modification to this approach, using a crystallizable interlayer, can give rise to 'Thermabond' joints that are not suceptible to attack even by reagents such as methylene dichloride; however, in such systems the processing window is significantly narrowed in comparison to amorphous interlayer bonding.

5.4 Rework, repair and reclaim

A particular advantage of thermoplastic composites is that they can be reprocessed many times and still develop the full inherent properties of the materials. This ability finds expression in the field of rework, repair and reclaim.

Ideally there should never be any need for rework during processing of composite materials. In practice there can be many causes for rework, including machine failure, failure of heaters, failure of ancillary equipment, and human error. Unfortunately the larger and more complex the component, the greater the possibility of some fault in the processing. With thermoplastic composites, a moulding that fails inspection for some reason can readily be reprocessed to give a quality component once the cause of the failure has been identified and corrected. Carlin[333] has used a 'worst case' reprocessing history for a high temperature semi-crystalline Aromatic Polymer Composite, and finds no deterioration in properties as a result.

The same arguments of reprocessability can be applied to a damaged structure[334–337]. Delaminated sheets, after repeated impacts, can be remoulded and the full original properties recovered[338]. Even mouldings that have suffered punch through failure can be remoulded to give a high proportion of the original performance[339]. Thermoplastic technology, and in particular 'Thermabond' (Section 5.3.5), provide a range of options for additive repair strategies. Tools for repair come readily to hand: a soldering iron is an excellent way of applying heat and pressure to a locally damaged structure.

Inevitably during processing there will be some offcuts of prepreg tape or trimmings of mouldings. Such materials can be chopped up and diluted with additional resin to provide high performance compounds for processing by injection moulding[340]. Prepreg tape offcuts, without dilution, can be consolidated into sheet or mouldings, to provide a stock material form whose properties are intermediate between injection moulding compounds and continuous fibre composites[341–343].

There is also potential to recover lifetime expired structures. Mouldings can be consolidated into thick blocks as a stock form for machining. Large area simple sheet structures can be recovered as a flat sheet stock for subsequent reprocessing into components by rapid stamping processes. Where no such higher value product can be established, the lifetime expired structures can be reprocessed in the same way as moulding trimmings, as a long fibre moulding compound of approximately one third the value of the original prepreg.

The potential for rework, repair and reclaim is a unique feature of thermoplastic composite materials that is likely to play an important role in determining total lifetime costs for composite structures.

5.5 Quality in processing

Quality in processing has many facets. The single most important fact to establish is that the final product is what it was designed to be.

With true thermoplastic systems, we know immediately that the chemistry is unchanging, provided that certain requirements of stability are met (Section 5.1.1 above). The greatest danger to this requirement is exposure to oxygen. Wherever possible, it is desirable to minimize contact with air at high temperature.

With semi-crystalline polymers, the crystallization characteristics are sensitive to processing history and to chemical structure (Section 4 above). It follows that a study of the crystallinity and crystallization behaviour of a small sample cut from a moulding is a particularly sensitive guide to its history. Since most mouldings will require some trimming, the inclusion of a small tab to be checked in this way offers a simple and effective verification of the status of the matrix.

With melt preimpregnated thermoplastics, each fibre is already fully wetted by the resin. The interface, on which the critical out of plane properties depend, is established, and there is no mechanism for it to be broken. This is not the case with any other product form (Section 3 above). The selection of the melt impregnated product form obviates any requirement to verify this characteristic.

Because of the high viscosity of thermoplastic polymers, once the resin/fibre distribution is established, there is little tendency to change. High pressures maintained for long times can cause some percolation of resin, which may be particularly noticeable at the ends of plies. A general recommendation is to carry out processing as fast as possible, but with as little pressure as is consistent with satisfactory forming.

During forming there may be a tendency for transverse flow of the fibres, leading to thinning of the plies. This is particularly noticeable in mouldings of complex form, where the forming pressure is applied first to high spots. Here the moulding tends to thin down, pushing fibres sideways into the valleys of the moulding. Transverse ply thinning can be as much as 75% of the original ply thickness. As with resin migration, it is preferred to form the component rapidly in order to minimize unwanted transverse flow. An alternative method of reducing this phenomenon is to use a woven product form, either directly impregnated woven fabric or woven tapes. The weave structure constrains local deformation of the material in respect of thickness variations. One result of transverse flow is that the line of the fibre varies from its original path. A more serious cause of distortion of the fibres is if the complex shaping geometry introduces local compressive stresses into the molten laminates. If these are not adequately supported, there can be a tendency to buckling. To avoid such buckling, the fibres should be constrained as much as possible during the process. Thus, in diaphragm forming, the use of stiff diaphragms, which more effectively constrain the prepreg, are to be preferred if buckling of the plies is a problem. Unfortunately such a course inevitably leads to high forming pressures and longer time scales of forming, and so more transverse flow. It should be possible to predict fibre movement during processing; at the present time it is a question of learning by experience and judging the optimum processing conditions. While today such fibre movement cannot be fully predicted, there is ample evidence that it is reproducible. Such fibre displacement can be readily monitored by the use of X-ray opaque tracer fibres.

The single most important aspect of the quality of a moulding is the level of consolidation. This is conveniently measured non-destructively by ultrasonic C-scan. The level of attenuation is sensitive to any discontinuities in the laminate, to the surface quality, and to the thickness of the component. Figure 5.36 shows the level of attenuation as a function of thickness for well consolidated laminates of carbon fibre reinforced PEEK. Most of the attenuation, which depends on the frequency used, occurs at the front and back faces, but there is also a small dependence on part thickness.

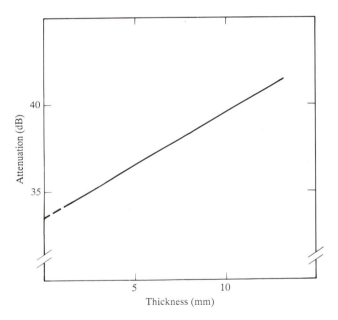

Figure 5.36 Ultrasonic C-scan attenuation as a function of specimen thickness for quasi-isotropic CF/PEEK panels containing 61 per cent by volume high strength carbon fibre. The data were obtained by using a Wells Krautkramer DSM35 flaw detector in the double pass reflectance mode

Quality in processing depends on reproducibility of the processing operation. While thermoplastic composite materials are not usually considered suitable for hand lay up, their suitability for machine controlled processes tends to eliminate one of the variables in the processing operation. The transition to automated processing, with fully impregnated, fully polymerized, thermoplastic product forms, potentially reduces the in-line quality control to the question 'What temperature did the material achieve during the processing cycle?'

References

5–1   F. N. COGSWELL, 'The Processing Science of Thermoplastic Structural Composites', *International Polymer Processing,* **1**, 4, pp. 157–165 (1987).

5–2   T. G. GUTOWSKI, 'A Resin Flow/Fiber Deformation Model for Composites', *SAMPE Quarterley,* **16**, pp. 58–64, (1985).

5–3   F. N. COGSWELL AND D. C. LEACH, 'Processing Science of Continuous Fibre Reinforced Thermoplastic Composites', *SAMPE J.,* **24**, 3, pp. 11–14 (1988).

5–4   W. SOLL AND T. GUTOWSKI, 'Forming Thermoplastic Composite Parts', 33rd International SAMPE Symposium, pp. 1004–1014, (1988).

5–5   C. L. TUCKER AND S. J. HWANG, *Process Modeling for Advanced Thermoplastic Matrix Composites: A Conceptual Outline,* University of Illinois (1987).

5–6   W. I. LEE AND G. S. SPRINGER, 'A Model of the Manufacturing Process of Thermoplastic Matrix Composites', *J. Composite Materials,* **21**, 11, pp. 1017–1055 (1987).

5–7   G. S. SPRINGER, 'Processing Model of Thermoplastic Matrix Composites', 33rd International SAMPE Symposium, pp. 661–669, (1988).

5–8   J. MUZZY, 'Processing of Advanced Thermoplastic Composites', ASME Conference, Atlanta, pp. 27–39 (1988).

5–9   J. MUZZY, L. NORPOTH AND B. VARUGHESE, 'Characterization of Thermoplastic Composites for Processing', *SAMPE J.,* **25**, 1, pp. 23–29 (1989).

5–10   F. N. COGSWELL, *Polymer Melt Rheology: a guide for industrial practice,* George Godwin (1981).

5–11   D. J. GROVES, 'A Characterization of Shear Flow in Continuous Fiber Thermoplastic Laminates', *Composites,* **20**, 1, pp. 28–32 (1989).

5–12   J. J. SCOBBO AND N. NAKAJIMA, 'Time–Temperature Processing Windows for Thermoplastic/ Graphite Fiber Composites', SPE ANTEC 90, pp. 1397–1400 (1990).

5–13   W. J. SICHINA AND P. S. GILL, 'Characterization of Composites by Thermal Analysis', 31st International SAMPE Symposium, pp. 1104–1112 (1986).

5–14   C.-C. M. MA, H.-C. HSIA, W.-L. LIU AND J.-T. HU, 'Thermal and Rheological Properties of Poly (Phenylene Sulphide) and Poly (Ether Etherketone) Resins and Composites', *Polymer Composites,* **8**, 4, pp. 256–264 (1987).

5–15   J. F. CARPENTER, 'Thermal Analysis and Crystallization Kinetics of High Temperature Thermoplastics', *SAMPE Journal,* **24**, 1, pp. 36–39 (1988).

5–16   See 5–14.

5–17   J. B. CATTANACH AND F. N. COGSWELL, 'Processing with Aromatic Polymer Composites' in *Developments in Reinforced Plastics – 5,* edited by G. Pritchard, pp. 1–38 Elsevier Applied Science Publishers (1986).

5–18   C. R. HART, E. NIELD AND J. B. ROSE, 'Polymer Composition', UK Patent 1,337,434 (1973).

5–19   See 5–1.

5–20   D. J. BLUNDELL AND F. M. WILLMOUTH, 'Crystalline Morphology of the Matrix of PEEK-Carbon Fibre Aromatic Polymer Composites Part 3: Prediction of Cooling Rates During Processing', *SAMPE Quarterly,* **17**, 2, pp. 50–58 (1986).

5–21   M. LEUNG, in 'Advances in Thermoplastic Matrix Composite Materials', edited by G. M. Newaz, ASTM STP 1044 (1989).

5–22   See 5–1.

5–23   See 5–20.

5–24   See 5–21.

5–25   S. M. GROVE, 'Thermal Modelling of Tape Laying with Continuous Carbon Fibre Reinforced Thermoplastics', *Composites,* **19**, 5, pp. 367–375 (1988).

5–26   R. J. DOWNS AND L. A. BERGLUND, *Apparatus for Preparing Thermoplastic Composites,* Linkoping Institute of Technology (1986).

5–27   C. N. VELISARIS AND J. C. SEFERIS, 'Heat Transfer Effects on the Processing–Structure Relationships of Polyetheretherketone (PEEK) Based Composites', *Science and Engineering of Composite Materials,* **1**, 1, pp. 13–22 (1988).

5–28   C. N. VELISARIS AND J. C. SEFERIS, 'Heat Transfer Effects on the Processing–Structure Relationships of Polyetheretherketone (PEEK) Based Composites', *Polym. Engng. and Sci.,* **28**, 9, pp. 583–591 (1988).

5–29   T. E. SALIBA, R. A. SERVAIS AND D. P. ANDERSON, 'Modelling of Heat Transfer During the Processing of Thermoplastic Composites', *Proc. Am. Soc. for Composites,* 3rd Technical Conf., pp. 459–467, Technomic Publishing (1988).

5–30   S. J. HWANG AND C. L. TUCKER, 'Heat Transfer Analysis of Continuous Fiber Thermoplastic Matrix Composites during Manufacture', *Journal Thermoplastic Composite Materials*, **3**, pp. 41–51 (1990).

5–31   See 5–27.

5–32   See 5–20.

5–33   See 5–26.

5–34   D. J. BLUNDELL, J. M. CHALMERS, M. W. MACKENZIE AND W. F. GASKIN, 'Morphology of Aromatic Polymer Composites Part I: Crystallinity', *SAMPE Quarterly*, **16** (4), p. 22–30 (1985).

5–35   See 5–27.

5–36   J. A. BARNES, I. J. SIMS, G. FARROW, P. JACKSON, G. WOSTENHOLM AND B. YATES, 'Thermal Expansion Behaviour of Thermoplastic Composite Materials', *J. Thermoplastic Composites Materials*, **3**, pp. 66–80 (1990).

5–37   S. H. OLSON, 'Manufacturing with Commingled Yarns, Fabrics and Powder Prepreg Thermoplastic Composite Materials', 35th International SAMPE Symposium, pp. 1306–1319 (1990).

5–38   J. F. PRATTE, W. H. KRUEGER AND I. Y. CHANG, 'High Performance Thermoplastic Composites with Poly (Ether Ketone Ketone) Matrix', 34th International SAMPE Symposium, pp. 2229–2242 (1989).

5–39   I. BREWSTER AND J. B. CATTANACH, 'Engineering with Long Fibre Thermoplastic Composites', SAMPE Conference, London, May (1983).

5–40   H. P. G. D'ARCY, *Les Fontaines Publique de la Ville de Dijon*, Dalmont (1856).

5–41   See 5–1.

5–42   See 5–2.

5–43   A. J. WHEELER, Ph.D thesis, University College of Wales, Aberystwyth (1990).

5–44   J. G. WILLIAMS, C. E. M MORRIS AND B. C. ENNIS, 'Liquid Flow Through Aligned Fiber Beds', *Polymer Engineering and Science*, **14**, 6, pp. 413–419 (1974).

5–45   R. DAVE, J. L. KARDOS AND M. P. DUDUKOVIC, 'A Model for Resin Flow During Composite Processing: Part 1 – General Mathematical Development', *Polymer Composites*, **8**, 1, pp. 29–38 (1987).

5–46   T. G. GUTOWSKI, Z. CAI, J. KINGERY AND S. J. WILLIAMS, 'Resin Flow/Fiber Deformation Experiments', *SAMPE Quarterly*, **17**, 4, pp. 54–58 (1986).

5–47   See 5–1.

5–48   See 5–43.

5–49   See 5–43.

5–50   R. C. LAM AND J. L. KARDOS, 'The Permeability of Aligned and Cross-Plied Fiber Beds during Processing of Continuous Fiber Composites', American Society for Composites, 3rd Annual Technical Conference, pp. 3–11 (1988).

5–51   See 5–1.

5–52   See 5–50.

5–53   See 5–50.

5–54   See 5–6.

5–55   See 5–10.

5–56   See 5–10.

5–57   See 5–10.

5–58   D. J. GROVES, private communication (1990).

5–59   See 5–2.

5–60   J. A. BARNES AND F. N. COGSWELL, 'Transverse Flow Processes in Continuous Fibre Reinforced Thermoplastics', *Composites*, **20**, 1, pp. 38–42 (1989).

5–61   F. N. COGSWELL AND D. J. GROVES, 'The Melt Rheology of Continuous Fibre Reinforced Structural Composite Materials', *Proc. Xth International Congr. Rheol*, Sydney (1988).

5–63   See 5–10.

5–64   See 5–11.

5–65   See 5–11.

5–66   See 5–61.

5–67   J. J. SCOBBO, J. R. AND N. NAKAJIMA, 'Dynamic Mechanical Analysis of Molten Thermoplastic/ Continuous Graphite Fiber Composites in Simple Shear Deformation', 21st International SAMPE Technical Conference, 25–28 September, pp. 730–743, (1989).

5–68   T. G. ROGERS, 'Rheological Characterization of Anisotropic Materials', *Composites*, **20**, 1, pp. 21–27 (1989).

5–69   D. J. GROVES, D. M. STOCKS AND A. M. BELLAMY, *Isotropic and Anisotropic Shear Flow in Continuous Fibre Thermoplastic Composites'*, British Society of Rheology, Edinburgh (1990).

5–70   C. CATHEY, University of Queensland, private communication (1989).

5–71   See 5–67.

5–72   See 5–67.

5–73   See 5–69.

5–74   See 5–70.

5–75   See 5–64.

5–76   See 5–61.

5–77   See 5–67.

5–78   See 5–69.

5–79   See 5–70.

5–80   See 5–10.

5–81   See 5–61.

5–82   See 5–61.

5–83   See 5–69.

5–84   See 5–69.

5–85   See 5–1.

5–86   See 5–60.

5–87   R. BALASUBRAMANYAM, R. S. JONES AND A. B. WHEELER, 'Modelling Transverse Flows of Reinforced Thermoplastic Materials', *Composites*, **20**, 1, pp. 33–37 (1989).

5–88   X. WU, 'Thermoforming of Thermoplastic Composites – Interply Shear Flow Analysis', 21st International SAMPE Technical Conference, pp. 915–923 (1989).

5–89   J. A. BARNES AND F. N. COGSWELL, 'Flow Processes in Continuous-Fibre Reinforced Thermoplastic Composites', American Society for Composites/University of Delaware Joint Symposium on Composite Materials Science and Engineering, Delaware, 23–25 September (1987).

5–90   J. A. BARNES, private communication (1990).

5–91   *Ibid.*

5–92   See 5–61.

5–93   See 5–1.

5–94   See 5–1.

5–95   See 5–17.

5–96   See 5–88.

5–97   P. V. KAPRIELIAN AND J. M. O'NEILL, 'Shearing Flow of Highly Anisotropic Laminated Composites', *Composites,* **20**, 1, pp. 43–47 (1989).

5–98   R. SCHERER, N. ZAHLAN AND K. FRIEDRICH, 'Modelling the Interply-slip Process During Thermoforming of Thermoplastic Composites Using Finite Element Analysis', 2nd International Conference on Computer Aided Design in Composite Materials Technology, CADCOMP 90, Brussels (1990).

5–99   T. G. ROGERS, Nottingham University, private communication (1989).

5–100 *Ibid.*

5–101 See 5–5.

5–102 See 5–6.

5–103 See 5–7.

5–104 See 5–8.

5–105 See 5–9.

5–106 See 5–20.

5–107 See 5–21.

5–108 See 5–25.

5–109 See 5–26.

5–110 See 5–27.

5–111 See 5–27.

5–112 See 5–29.

5–113 G. R. BELBIN, I. BREWSTER, F. N. GOGSWELL, D. J. HEZZELL AND M. S. SNERDLON, 'Carbon Fibre Reinforced PEEK: a Thermoplastic composite for Aerospace Applications', European SAMPE at Stresa (1982)

5–114 See 5–39.

5–115 See 5–17.

5–116 See 5–17.

5–117 See 5–39.

5–118 J. B. CATTANACH, G. CUFF AND F. N. COGSWELL, 'The Processing of Thermoplastics Containing High Loadings of Long and Continuous Reinforcing Fibres', *Journal of Polymer Engineering*, **6**, 1–4, pp. 345–362 (1986).

5–119 A. DUTHIE, 'Engineering Substantiation of Fibre Reinforced Thermoplastic Composites for Aerospace Primary Structure', 33rd International SAMPE Symposium, pp. 296–307 (1988).

5–120 J. D. MUZZY, *Processing of Advanced Thermoplastic Composites*, Georgia Institute of Technology (1988).

5–121 J. D. MUZZY, X. WU AND J. S. COLTON, 'Thermoforming of High Performance Thermoplastics Composites', ANTEC, pp. 1465–1470 (1989).

5–122 R. K. OKINE, 'Analysis of Forming Parts from Advanced Thermoplastic Composite Sheet Materials', *SAMPE Journal*, **25**, 3, pp. 9–19 (1989).

5–123 D. A. ARCHIBALD, J. W. SCHWARZ AND J. L. WANAMAKER, 'Reducing the Cost of Thermoplastic Composite Structures', *Proc. Am. Soc. for Composites, 4th Tech. Conference*, pp. 593–601, Technomic (1989).

5–124 M. FOLEY AND E. BERNARDON, 'Cost Estimation Techniques for the Design of Cost Effective Automated Systems for Manufacturing Thermoplastic Composite Structures', 35th International SAMPE Symposium, pp. 1321–1335 (1990).

5–125 J. SHUKLA, 'Fabrication of Aircraft Structures from Thermoplastic Drapeable Preforms', 21st International SAMPE Technical Conference, pp. 700–704 (1989).

5–126 See 5–119.

5–127 See 5–123.

5–128 See 5–124.

5–129 See 5–6.

5–130 G. SPRINGER, Stanford University, private communication (1990).

5–131 See 5–37.

5–132 J. D. RUSSELL, 'Comparison of Processing Techniques for Filmix Unidirectional Commingled Fabric', 35th International SAMPE Symposium, pp. 1118–1130 (1990).

5–133 M. THIEDE-SMET, 'Study of Processing Parameters of PEEK/Graphite Composite Fabricated with 'FIT' Prepreg', 34th International SAMPE Symposium, pp. 1265–1274 (1989).

5–134 See 5–17.

5–135 See 5–1.

5–136 See 5–60.

5–137 See 5–1.

5–138 See 5–7.

5–139 See 5–26.

5–140 G. KEMPE AND H. KRAUSS, 'Processing and Mechanical Properties of Fiber-Reinforced Polyetheretherketone (PEEK)', 14th Congress of the International Council of the Aeronautical Sciences, Toulouse (1984).

5–141 G. KEMPE AND H. KRAUSS, 'Manufacturing Processes and Moulding of Fibre-Reinforced Polyetheretherketone (PEEK), ICAS-86–4.6.1 (1986).

5–142 J. C. SEFERIS, 'Polyetheretherketone (PEEK): Processing, Structure and Properties Studies for a Matrix in High Performance Composites', *Polymer Composites*, **7**, p. 158 (1986).

5–143 J. M. IACONIS, 'Process Variables Evaluation of PEEK APC-2 Thermoplastic Matrix Composite', 32nd International SAMPE Symposium, (1987).

5–144 D. P. SINGH, O. VERBERNE, 'Fabrication Optimization of Continuous Fibre Reinforced Thermoplastic Composites', SAMPE Conference, La Baule (1987).

5–145 A. J. SMILEY, R. B. PIPES, 'Simulation of the Diaphragm Forming of Carbon Fibre/Thermoplastic Composite Laminates', American Society for Composites, 2nd Technical Conference, Delaware (1987).

5–146 J. C. SEFERIS, C. AHLSTROM, S. H. DILLMAN, 'Cooling Rate and Annealing as Processed Parameters for Semi-Crystalline Thermoplastic-Based Composites', SPE ANTEC '87, pp. 1467–1470 (1987).

5–147 L. NORPOTH, A. BUTT AND J. MUZZY, 'Quantitative Analysis of APC-2 Consolidation', 33rd International SAMPE Symposium, pp. 1331–1341 (1988).

5–148 S. GROSSMAN, M. AMATEAU, 'The Effect of Processing in a Graphite Fibre/Polyetheretherketone Thermoplastic Composite', 33rd International SAMPE Symposium, pp. 681–692 (1988).

5–149 B. R. COXON, J. C. SEFERIS, L. B. ILCEWICZ, 'The Effects of Process Variables on Transverse Matrix Cracking in High Performance Composites', *Process Manufacturing International*, **4**, pp. 129–140 (1988).

5–150 Y. SAKATANI, M. YOSHIDA AND Y. YAMAGUCHI, *Fabrication Studies in Carbon/Thermoplastic Matrix Composites,* Mitsubishi Heavy Industries, Nagoya Aircraft Works, ACCE (1988).

5–151 T. W. KIM, E. J. JUN AND W. I. LEE, 'The Effect of Pressure on the Impregnation of Fibers with Thermoplastic Resins', 34th International SAMPE Symposium, pp. 323–328 (1989).

5–152 B. J. ANDERSON AND J. S. COLTON, 'A Study in the Lay-Up and Consolidation of High Performance Thermoplastic Composites', *SAMPE Journal,* **25**, 5, pp. 22–28 (1989).

5–153 E. CORRIGAN, D. C. LEACH AND T. McDANIELS, 'The Influence of Processing Conditions on the Properties of PEEK Matrix Composites', 10th Int. Meeting, Eur. Chapter SAMPE, Materials Science Monographs, **55**, pp. 121–132, Elsevier (1989).

5–154 J. A. E. MANSON, T. L. SCHNEIDER AND J. C. SEFERIS, 'Press Forming of Continuous Fiber Reinforced Thermoplastic Composites', *Polymer Composites* (in press).

5–155 J. A. E. MANSON AND J. C. SEFERIS, 'Autoclave Processing of PEEK/Carbon Fiber Composites', *J. Thermoplastic Composite Materials,* **2**, pp. 34–42 (1989).

5–156 J. A. E. MANSON AND J. C. SEFERIS, 'Process Analysis and Properties of High Performance Thermoplastic Composites, in *Engineering. Applications of New Composites,* edited by S. A. Paipetis and G. C. Papanicolaou, Omega Scientific (1988).

5–157 'Fabricating with Aromatic Polymer Composite APC-2', Data Sheet No 5, Fiberite Corporation (1986).

5–158 See 5–140.

5–159 T. NAGUMO, H. NAKAMURA, Y. YOSHIDA AND K. HIRAOKA, 'Evaluation of PEEK Matrix Composites', 32nd International SAMPE Symposium, pp. 486–497 (1987).

5–160 See 5–143.

5–161 See 5–156.

5–162 See 5–140.

5–163 See 5–60.

5–164 See 5–60.

5–165 See 5–140.

5–166 See 5–130.

5–167 See 5–154.

5–168 See 5–140.

5–169 See 5–157.

5–170 D. GUPTA, 'High Temperature Composite Tooling for Advanced Composite Manufacture', in *Materials and Processing – Move into the 90's, Proc. 10th Int. Eur. Chapter Conf. of SAMPE.,* Materials Science Monographs, **55**, pp. 293–302, Elsevier (1989).

5–171 A. KERR, 'Integrally Heated Tooling for Economical, non-Autoclave Production of Thermoplastic Parts', 35th International SAMPE Symposium, pp. 1917–1927 (1990).

5–172 *Ibid.*

5–173 See 5–17.

5–174 See 5–39.

5–175 P. J. IVES, 'Thermoplastic Composites: The First Step Towards a Continuous Production Method', *Proc. PRI Conference 'Automated Composites '88',* Leeuwenhorst, 21/1–21/10 (1988).

5–176 J. S. KENWORTHY AND J. B. CATTANACH, ICI Composite Structures Group, private communication.

5–177 See 5–9.

5–178 See 5–152.

5–179 See 5–152.

5–180 F. N. COGSWELL, D. J. HEZZELL AND P. J. WILLIAMS, 'Fibre Reinforced Composites and Methods for Producing such Composites', USP 4559262 (1981).

5–181 G. COWEN, U. MEASURIA AND R. M. TURNER, *Section Pultrusion of Continuous Fibre Reinforced Thermoplastics,* Institute of Mechanical Engineers, C22/86, 105–112 (1986).

5–182 L. H. PROVEROMO, W. K. MEUNCH, W. MARX AND G. LUBIN, *SAMPE Journal,* **17**, 1, pp. 10–11 (1981).

5–183 J. E. O'CONNOR, 'Polyphenylene Sulphide Pultruded Type Composite Structures', 42nd Annual Conference, Reinforced Plastics, Composites Institute, Society of Plastics Industry, 1-D (1987).

5–184 M. L. WILSON, I. O. MACCONOCHIE AND G. S. JOHNSON, 'On-Orbit Fabrication of Large Space Structures Using Thermoplastic Preimpregnated Graphite Tapes and a Pultrusion Process', 43rd Annual Conference, Composite Institute, Society of Plastics Industry 6B, pp. 1–7 (1988).

5–185 J. COLTON, J. LYONS, B. LUKASIC, J. MAYER, S. WITTE AND J. MUZZY, 'On-Orbit Fabrication of Space Station Structures', 34th International SAMPE Symposium, pp. 810–816 (1989).

5–186 See 5–182.

5–187 H. J. GYSIN, 'Advanced Thermoplastic Matrix Composites in Mechanical Applications', *Proc. Fibre Reinforced Composites FRC '90*, Institute of Mechanical Engineers, pp. 69–72 (1990).

5–188 See 5–39.

5–189 S. M. GROVE, 'Thermal Modelling of Tape Laying with Continuous Carbon Fibre-Reinforced Thermoplastic', *Composites*, **19**, 5, pp. 367–375 (1988).

5–190 E. P. BEYELER AND S. I. GUCERI, 'Thermal Analysis of Laser – Assisted Thermoplastics – Matrix Composite Tape Consolidation', Transactions of the ASME, *Journal of Heat Transfer*, **110**, pp. 424–430 (1988).

5–191 E. P. BEYELER, W. PHILLIPS AND S. I. GUCERI, 'Experimental Investigation of Laser Assisted Thermoplastic Tape Consolidation', *J. Thermoplastic Composites*, **1**, pp. 107–121 (1988)

5–192 See 5–152.

5–193 S. M. GROVE AND D. SHORT, 'Evaluation of Carbon Fibre Reinforced PEEK Composites Manufactured by Continuous Local Welding of Prepreg Tape', *Plast. Rubber Process Appl.*, **10**, 1, pp. 35–44 (1988).

5–194 See 5–152.

5–195 See 5–193.

5–196 G. GRIFFITHS, W. HILLIER AND J. WHITING, 'Thermoplastic Composite Manufacturing Technology for a Flight Standard Tailplane', 33rd International SAMPE Symposium, pp. 308–316 (1988).

5–197 D. H. BOWEN, 'Filament Winding in the 1980's', *Fibre Reinforced Composites '84*, Liverpool (1984).

5–198 M MAHLKE, 'Processing of Fibre Reinforced Thermoplastics', *Engineering Plastics*, **2**, 3, pp. 163–195 (1989).

5–199 M. EGERTON AND M. GRUBER, 'Thermoplastic Filament Winding Demonstrating Economics and Properties Via In-Situ Consolidation', 33rd International SAMPE Symposium, pp. 35–46 (1988).

5–200 S. GUCERI, University of Delaware, private communication (1988).

5–201 O. DICKMAN, K. LINDERSSON AND L. SVENSSON, 'Filament Winding of Thermoplastic Matrix Composites', submitted to *Journal of Plastics and Rubber Processing and Applications* (1989).

5–202 J. H. C. ROWAN AND R. N. ASKANDER, 'Filament Winding of High Performance Thermoplastic Composites', in *Materials and Processing – Move into the 90's*, *Proc. 10th Int. Eur. Chapter Conf. of SAMPE*, Materials Science Monographs, **55**, pp. 11–20, Elsevier (1989).

5–203 See 5–199.

5–204 See 5–200.

5–205 See 5–202.

5–206 See 5–201.

5–207 M. CIRINO, Ph.D Thesis, University of Delaware (1989).

5–208 A. BRAGE AND C. LAMBRELL, 'Heat Flow Analysis in Connection with Thermoplastic Filament Winding', *SAMPE Quarterly*, **10**, 3, pp. 31–35 (1988).

5–209 G. M. WELLS AND K. F. MCANULTY, 'Computer Aided Filament Winding Using Non-Geodesic Trajectories', ICCM-6, **1**, pp. 161–173 (1987).

5–210 See 5–17.

5–211 See 5–17.

5–212 See 5–140.

5–213 See 5–15.

5–214 See 5–17.

5–215 See 5–17.

5–216 See 5–39.

5–217 See 5–144.

5–218 See 5–196.

5–219 G. R. GRIFFITHS, J. W. DAMON AND T. T. LAWSON, 'Manufacturing Techniques for Thermoplastic Matrix Composites', *SAMPE Journal*, **21**, 6, pp. 32–35 (1984).

5–220 G. R. GRIFFITHS, W. D. HILLIER AND J. A. S. WHITING, 'Thermoplastic Composite Manufacturing Technology for a Flight Standard Tailplane', *SAMPE Journal*, **25**, 3, pp. 29–34 (1989).

5–221 See 5–196.

5–222 A. J. BARNES AND J. B. CATTANACH, 'Advances in Thermoplastic Composite Fabrication Technology', Materials Engineering Conference, Leeds (1985).

5–223 P. J. MALLON AND C. M. O'BRADAIGH, 'Development of a Pilot Autoclave for Polymeric Diaphragm Forming of Continuous Fibre-Reinforced Thermoplastics', *Composites*, **19**, 1, pp. 37–47 (1988).

5–224 A. J. SMILEY AND R. B. PIPES, 'Simulation of the Diaphragm Forming of Carbon Fibre/Thermoplastic Composite Laminates', American Society for Composites, 2nd Technical Conference, in Delaware (1987).

5–225 C. M. O'BRADAIGH AND P. J. MALLON, 'Effect of Forming Temperature on the Properties of Polymeric Diaphragm Formed APC-2 Components', American Society for Composites, 2nd Conference, Delaware (1987).

5–226 C. M. O'BRADAIGH AND P. J. MALLON, 'Effect of Forming Temperature on the Properties of Polymer Diaphragm Formed Components', ASCM/CCM Joint Symposium on Composite Materials Science and Engineering (1987).

5–227 P. J. MALLON, C. M. O'BRADAIGH AND R. B. PIPES, 'Polymeric Diaphragm Forming of Continuous Fiber Reinforced Thermoplastics', 33rd International SAMPE Symposium, pp. 47–61 (1988).

5–228 P. J. MALLON, C. M. O'BRADAIGH AND R. B. PIPES, 'Polymeric Diaphragm Forming of Complex-Curvature Thermoplastic Composite Parts', submitted to Composites (1988).

5–229 P. J. MALLON, C. M. O'BRADAIGH AND R. B. PIPES, 'Polymeric Diaphragm Forming of Continuous Fibre Reinforced Thermoplastic Matrix Composites', Composites, 20, 1, pp. 48–56 (1989).

5–230 P. J. MALLON AND C. M. O'BRADAIGH, 'Development of a Pilot Autoclave for Polymeric Diaphragm Forming of Continuous Fibre Reinforced Thermoplastics', Composites, 19, 1, pp. 37–48 (1988).

5–231 C. M. O'BRADAIGH, M. F. FLEMING, P. J. MALLON AND R. B. PIPES, 'Effects of Forming Rate on Polymeric Diaphragm Forming of Thermoplastic Composites', Proc. PRI Conference 'Automated Composites '88', Leeuwenhorst, 23/1–23/8 (1988).

5–232 A. J. SMILEY AND R. B. PIPES, 'Analysis of the Diaphragm Forming of Continuous Fiber Reinforced Thermoplastics', Journal of Thermoplastic Composite Materials, 1, pp. 298–321 (1988).

5–233 M. N. WATSON, Techniques for Joining Thermoplastic Composites', The Welding Institute, Abington, Cambridge (1989).

5–234 M. R. MONAGHAN, C. M. O'BRADAIGH, P. J. MALLON AND R. B. PIPES, 'The Effect of Diaphragm Stiffness on the Quality of Diaphragm Formed Thermoplastic Composite Components', 35th International SAMPE Symposium, pp. 810–828 (1990).

5–235 C. M. O'BRADAIGH, R. B. PIPES AND P. J. MALLON, 'Issues in Diaphragm Forming of Continuous Fibre Reinforced Thermoplastics', submitted to Polymer Composites (1990).

5–236 M. R. MONAGHAN AND P. J. MALLON, 'Development of a Computer Controlled Autoclave for Forming Thermoplastic Composites', Composite Manufacturing, 1, 1, pp. 8–14 (1990).

5–237 R. B. OSTROM, S. B. KOCH AND D. L. WIRZ-SAFRANEK, 'Thermoplastic Composite Fighter Forward Fuselage', SAMPE Quarterly, 21, 1, pp. 39–45 (1989).

5–238 See 5–222.

5–239 See 5–229.

5–240 See 5–231.

5–241 See 5–235.

5–242 See 5–230.

5–243 See 5–232.

5–244 See 5–235.

5–245 See 5–17.

5–246 A. B. STRONG AND P. HAUWILLER, 'Incremental Forming of Large Fiber-Reinforced Thermoplastic Composites', 34th International SAMPE Symposium, pp. 43–54 (1989).

5–247 H. SAATCHI, 'Isoclave Isostatic In-Situ Fabrication of Thermoplastic Composite Sandwich Structure', 35th International SAMPE Symposium, pp. 245–255 (1990).

5–248 CATTANACH AND COGSWELL, op. cit.

5–249 T. STEVENS, 'Joining Advanced Thermoplastic Composites', Composite Materials, 107, 3, pp. 41–45 Penton (1990).

5–250 D. MAGUIRE, 'Joining Thermoplastic Composites', SAMPE Journal, 25, 1, pp. 11–14 (1989).

5–251 R. WALSH, M. VEDULA AND M. J. KOCZAK, 'Comparative Assessment of Bolted Joints in a Graphite Reinforced Thermoset vs Thermoplastic', SAMPE Quarterly, 20, 4, pp. 15–19 (1989).

5–252 R. E. POLITI, 'Factors Affecting the Performance of Composite Bonded Structure', 19th International SAMPE Technical Conference (1987).

5–253 A. J. KINLOCH AND C. M. TAIG, 'The Adhesive Bonding of Thermoplastic Composite', Journal of Adhesion, 21, 291–302 (1987).

5–254 R. GREER, 'Structural Adhesive Bonding of Carbon Fiber Reinforced Thermoplastic Composites', 33rd International SAMPE Symposium (1988).

5–255 See 5–250.

5–256 s. y. wu, a. m. schuler and d. n. keane, 'Adhesive Bonding of Thermoplastic Composites 1: Effect of Surface Treatment on Adhesive Bonding', 19th International SAMPE Technical Conference, pp. 277–290 (1987).

5–257 g. k. a. kodakian and a. j. kinloch, 'Structural Adhesive Bonding of Thermoplastic Fibre Composite', Bonding and Repair of Composites Seminar organized jointly by Rapra Technology and Butterworth Scientific Limited, Birmingham (1989).

5–258 j. w. powers and w. j. trzaskos, 'Recent Developments in Adhesives for Bonding Advanced Thermoplastic Composites', 34th International SAMPE Symposium, pp. 1987–1998 (1989).

5–259 d. m. maguire, 'Joining Thermoplastic Composites', Composite Polymers, 2, 4, pp. 325–334 (1989).

5–260 d. k. kohli, 'Development of Improved 121°C (250°F) Cure Adhesives for Aerospace Applications – FM 300-2 Adhesive System' in Materials and Processing – Move into the 90's, Proc. 10th Int. Eur. Chapter Conf. of SAMPE, Materials Science Monographs, 55, pp. 239–250, Elsevier (1989).

5–261 s.-i. y. wu, 'Adhesive Binding of Thermoplastic Composites 1: The Effect of Surface Treatments on Adhesive Bonding', 19th International SAMPE Technical Conference, pp. 277–286 (1987).

5–262 s.-i. y. wu, 'Adhesive Bonding of Thermoplastic Composites 2: Surface Treatment Study', 35th International SAMPE Symposium, pp. 846–858 (1990).

5–263 See 5–252.

5–264 See 5–256.

5–265 See 5–253.

5–266 See 5–256.

5–267 See 5–257.

5–268 See 5–256.

5–269 See 5–257.

5–270 See 5–257.

5–271 See 5–261.

5–272 See 5–261.

5–273 b. a. stein, w. t. hodges and j. r. tyeryar, 'Rapid Adhesive Bonding of Advanced Composites and Titanium', presented at the AIAA/ASME/ASCE/AHS 26th Structures, Structural Dynamics and Materials Conference, Orlando (1985).

5–274 e. nield, Private communication (1983).

5–275 See 5–224.

5–276 See 5–250.

5–277 See 5–259.

5–278 a. benatar and t. g. gutowski, 'A Review of Methods for Fusion Bonding Thermoplastic Composites', SAMPE Journal, 23, 1, pp. 33–39 (1987).

5–279 w. krenkel, 'Joining Techniques for CFR Polyether Etherketone (PEEK)', Joint Israeli-German Status Seminar on Structure Performance of Polymer Composites, Tel Aviv (1987).

5–280 n. a. taylor, 'The Feasibility of Welding Thermoplastic Composite Materials', Bonding and Repair of Composites Seminar organized jointly by Rapra Technology and Butterworth Scientific Limited (1989).

5–281 n. a. taylor, 'The Feasibility of Welding APC-2 Thermoplastic Composite Materials', The Welding Institute Research Bulletin, pp. 221–229 (1987).

5–282 e. m. silverman and r. a. criese, 'Joining Methods for Graphite/PEEK Thermoplastic Composites', SAMPE Journal, 25, 5, pp. 34–38 (1989).

5–283 See 5–157.

5–284 See 5–233.

5–285 a. benatar and t. gutowski, 'Ultrasonic Welding of Advanced Thermoplastic Composites', 33rd International SAMPE Symposium, pp. 1787–1797 (1988).

5–286 a. m. maffezzoli, j. m. kenny and l. nicolais, 'Welding of PEEK/Carbon Fiber Composite Laminates', SAMPE Journal, 25, 1, pp. 35–39 (1989).

5–287 j. border and r. salas, 'Induction Heated Joining of Thermoplastic Composites without Metal Susceptors', 34th International SAMPE Symposium, pp. 2569–2578 (1989).

5–288 e. c. eveno and j. w. gillespie, 'Resistance Welding of Graphite Polyetheretherketone Composites: An Experimental Investigation', Journal of Thermoplastic Composite Material, 1, pp. 322–338, October (1988).

5–289 r. c. don, l. bastien, t. b. jakobsen and j. w. gillespie, 'Fusion Bonding of Thermoplastic Composites by Resistance Heating', SAMPE Journal, 26, 1, pp. 59–66 (1990).

5–290 G. KEMPE AND H. KRAUSS, 'Moulding and Joining of Continuous Fiber-Reinforced Polyetheretherketone (PEEK)', *Proc. 16th Inter. Council Aeron. Science* (ICAS), **2**, pp. 1789–1800 (1988).

5–291 A. STRONG, D. P. JOHNSON AND B. A. JOHNSON, 'Variables Interactions in Ultrasonic Welding of Thermoplastic Composites', *SAMPE Quarterly,* **21**, 2, pp. 36–41 (1990).

5–292 G. WILLIAMS, S. GREEN, J. MCAFEE AND C. M. HEWARD, *Induction Welding of Thermoplastic Composites',* Institute of Mechanical Engineers, C400/034, pp. 133–136 (1990).

5–293 L. BASTIEN, R. C. DON AND J. W. GILLESPIE, JR., 'Processing and Performance of Resistance Welded Thermoplastic Composites', *Proc. 45th SPI Annual Conference,* Washington DC, 20-B, February 12–16 (1990).

5–294 See 5–250.

5–295 See 5–278.

5–296 See 5–286.

5–297 See 5–280.

5–298 See 5–288.

5–299 See 5–289.

5–300 See 5–293.

5–301 See 5–278.

5–302 See 5–250.

5–303 See 5–259.

5–304 See 5–280.

5–305 See 5–290.

5–306 See 5–292.

5–307 See 5–278.

5–308 See 5–278.

5–309 See 5–250.

5–310 See 5–259.

5–311 See 5–278.

5–312 See 5–280.

5–313 See 5–290.

5–314 See 5–291.

5–315 See 5–278.

5–316 See 5–278.

5–317 See 5–280.

5–318 See 5–288.

5–319 C. G. BOOTH, private communication (1989).

5–320 J. GILLESPIE, private communication (1990).

5–321 J. B. CATTANACH, private communication (1983).

5–322 C. BOOTH, A. J. SMILEY, M. T. HARVEY, F. N. COGSWELL AND P. J. MEAKIN, 'Thermoplastic Interlayer Bonding of Aromatic Polymer Composites', Bonding and Repair of Composites Seminar organized jointly by Rapra Technology and Butterworth Scientific Limited, pp. 39–44 July (1989).

5–323 F. N. COGSWELL, P. J. MEAKIN, A. J. SMILEY, M. T. HARVEY AND C. BOOTH, 'Thermoplastic Interlayer Bonding of Aromatic Polymer Composites', 34th International AMPE Symposium, pp. 2315–2325 (1989).

5–324 F. N. COGSWELL AND P. J. MEAKIN, 'Fibre Reinforced Thermoplastic Composite Structures', UK patent application 8728887 (1987).

5–325 See 5–322.

5–326 A. J. SMILEY AND A. HALBRITTER, 'Dual Polymer Bonding for Thermoplastic Composite Structures', SPE Antec (1990).

5–327 L. BASTIEN, J. W. GILLESPIE AND C. T. LAMBING, 'Strength of Semicrystalline Thermoplastic Composite Joints Using Dual Film Technology', American Society for Composites, 5th Technical Conference (1990).

5–328 See 5–289.

5–329 See 5–293.

5–330 See 5–282.

5–331 See 5–326.

5–332 P. J. MEAKIN, private communication (1990).

5–333 D. M. CARLIN, W. A. PRICE AND D. P. ANDERSON, 'Reprocessability of APC (ITX) Composite Material', 35th International SAMPE Symposium, pp. 46–58 (1990).

5–334 F. N. COGSWELL, 'Continuous Fibre Reinforced Thermoplastics', in *Mechanical Properties of Reinforced Thermoplastics'*, edited by D. W. Clegg and A. A. Collyer, pp. 83–118, Elsevier (1986).

5–335 E. A. WESTERMAN AND P. E. ROLL, 'An Apparatus to Prepare Composites for Repair', 34th International SAMPE Symposium, pp. 1041–1051 (1989).

5–336 C. L. ONG, M. F. SHEU, Y. Y. L. C. SHAN, 'The Repair of Thermoplastic Composites After Impact', 34th International SAMPE Symposium, pp. 458–469 (1989).

5–337 W. J. CANTWELL, P. DAVIES, P.-Y. JAR, P.-E. BOURBAN AND H. H. KAUSCH, 'Joining and Repair of Carbon Fiber Composites', *Proc. 11th International European Chapter SAMPE Conference,* pp. 411–426 (1990).

5–338 See 5–336.

5–339 See 5–336.

5–340 See 5–334.

5–341 G. C. MCGRATH, D. W. CLEGG, A. A. COLLYER AND U. MEASURIA, 'The Mechanical Properties of Compression Moulded Reconstituted Carbon Fibre Reinforced PEEK (APC-2)', *Composites,* **19**, 3, pp. 211–216 (1988).

5–342 D. W. CLEGG, G. MCGRATH AND M. MORRIS, 'The Properties and Economics of Reclaimed Long Fibre Thermoplastic Composites', *Composites Manufacturing,* **1**, 2, pp. 85–89 (1990).

5–343 I. K. PARTRIDGE, D. J. STEPHENSON AND K. M. AL-TURK, 'Recycling Waste APC', ECCM-4 (1990).

6 Dimensional stability

The first function of a structural material is to provide a state of architectural equilibrium. This requirement demands that the presence of applied load, either mechanical, thermal or dilational, shall not deform the component in such a way as to impair the performance of the structure. In the case of fibre reinforced composite materials we must be conscious of their anisotropic character, so as to ensure a full description of the material that can be used for design purposes. This is, of course, a complexity in comparison with isotropic materials, but, it also defines the opportunity to tailor the anisotropic properties of the material into the design of the structure to achieve optimum advantage from the stiff reinforcing fibres[1].

6.1 Intumescence

Most organic polymers will take up water to some extent. This tendency is particularly marked in certain classes of polymers such as polyamides and epoxy resins, where, through hydrogen bonding, the water can be chemically complexed into the structure. As we shall note in Chapter 10, polyetheretherketone is insensitive to virtually all liquid reagents and to water in particular. Carbon fibre reinforced PEEK composite will take up about 0.25% by weight of water after prolonged exposure at high temperature[2,3], but no dimensional change resulting from this has been detected, and it is assumed that the water is contained mainly in the free volume[4] and does not cause any significant swelling of the material.

One liquid, encountered in normal service, does appear to plasticize the matrix in PEEK based composites. This swelling agent is aviation fuel[5,6]. This phenomenon, which only occurs in the presence of stress, is considered in Chapter 10. The maximum fluid uptake is noted to be about 0.8%, and the only significant swelling that takes place is in the thickness of the structure, transverse to the plane of the fibres. The length of the component, the critical factor in determining the structural architecture, is unchanged. Since a proportion of the fluid will be taken up by free volume in the matrix, the maximum dimensional change in carbon fibre PEEK composite associated with such swelling is likely to be less than a 0.5% increase in thickness.

6.2 Thermal expansion

The coefficient of linear thermal expansion of polyetheretherketone is moderate and positive, and typical of most thermoplastic polymers. Across the diameter of the fibre the thermal expansion coefficient of carbon fibres is somewhat lower than that of the matrix, but, along the fibre direction, the expansivity is small and usually negative. This extreme difference in properties makes thermal expansion a strongly anisotropic property (Table 6.1).

Table 6.1 Thermal expansion of carbon fibre/PEEK at 23°C

	Coefficient of linear thermal expansion at 23°C (/°C)
Constituents	
PEEK resin	47.0×10^{-6}
High strength carbon fibre (axial)	-1.2×10^{-6}
High strength carbon fibre (transverse)	12.0×10^{-6}
61% by volume CF/PEEK composite	
Uniaxial laminate (axial), α_1	0.2×10^{-6}
Uniaxial laminate (transverse in plane), α_2	28.8×10^{-6}
Uniaxial laminate (through thickness), α_3	$(29.2 \times 10^{-6})*$
Cross plied laminate (in plane)	7.5×10^{-6}
Cross plied laminate (through thickness)	$(38.0 \times 10^{-6})*$
Quasi-isotropic laminate (in plane)	2.9×10^{-6}
Quasi-isotropic laminate (through thickness)	$(48.0 \times 10^{-6})*$

* Estimated values (see Appendix 14).

Barnes[7] presents a comprehensive report of the thermal expansion of carbon fibre reinforced polyetheretherketone in the temperature range $-180°C$ to $+120°C$. Figures 6.1 and 6.2 show the axial and transverse thermal expansion behaviour of a composite based on 61% high strength fibres in this temperature range: it is important to note the twentyfold difference in the scale of the vertical axes of the two graphs.

The transverse thermal expansion coefficient, α_2, reported here is a transverse measurement in the plane of the laminate. The coefficient of thermal expansion through the thickness, α_3, should be slightly higher, because of the higher thermal expansion of the thin resin rich interlayers. Appendix 14 suggests a through-thickness thermal expansion that is about 1 to 2% higher than the transverse in plane value, but for practical applications that difference can be ignored.

The thermal expansion of carbon fibre composite materials is low in comparison with most metals. Such materials can be tailored to provide near zero thermal expansion over a wide temperature range. This makes composites highly desirable from the standpoint of dimensional stability where thermal excursions are anticipated.

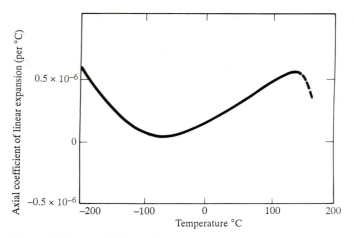

Figure 6.1 Axial coefficient of linear thermal expansion of CF/PEEK, based on 61% by volume of high strength carbon fibre, α_1

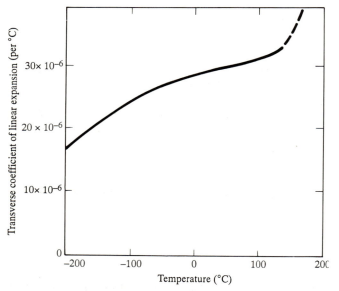

Figure 6.2 Transverse coefficient of linear thermal expansion of CF/PEEK, based on 61% by volume of high strength carbon fibres, α_2 (Appendix 8)

No less important than low thermal expansion coefficient is the possibility of permanent deformation, hysteresis, after a thermal cycle. There can be a slight change in dimensions after a first thermal cycle, and this can be attributed to the release of constraints applied during the fabrication process. In some composite materials the stresses induced by repeated thermal cycling can lead to microcracking and a progressive change in properties. A tough thermoplastic matrix inhibits such cracking, and Barnes[8] demonstrates that 100 cycles from

$-160°C$ to $+120°C$ produce no measurable change in the thermal expansion coefficient of carbon fibre/PEEK composites.

The anisotropy of the coefficient of linear thermal expansion gives the opportunity to tailor this property by selecting the fibre orientation. In particular it is possible to design a laminate to have a zero coefficient of thermal expansion in a preferred direction (Figure 6.3).

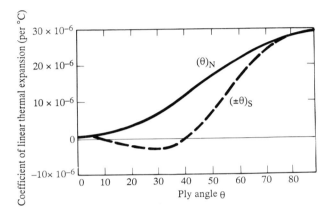

Figure 6.3 Influence of ply angle on coefficient of linear thermal expansion for CF/PEEK, based on 61% by volume high strength carbon fibres at 23°C, following results of Barnes[9]

Note, in particular, that for laminates of $\pm 6°$ and $\pm 40°$ the linear thermal expansion is zero. This result is of course only true at one temperature. The optimum angle of plied laminate for minimum thermal expansion depends on the thermal expansion along and across the fibres. These depend on temperature. Classical laminate theory[10] can be used to predict the effect of ply lay up for different temperatures, and so the optimum laminate to minimize distortion for any defined thermal excursion.

6.3 Stiffness

The primary reason for incorporating high loadings of collimated, continuous, reinforcing fibres into matrix resins is to enhance stiffness. Where structures are designed so that the fibres are oriented along the natural load paths, composites achieve their most efficient expression. Real structures, in real service conditions, must withstand off axis and out of plane loads. As well as measuring the optimum tensile stiffness of the material along the fibre direction, we must also study its response to loads at an angle to the fibre direction. Shear or torsional loadings, as well as simple elongation, must also be considered. Practical service also requires resistance to stress for a protracted period: creep resistance is a significant feature in predicting long term dimensional stability.

6.3.1 Stiffness characterization of uniaxial laminates

In a simple uniaxial laminate there are fifteen stiffness characteristics: three tensile moduli, six Poisson's ratios, and six shear moduli (Figure 6.4).

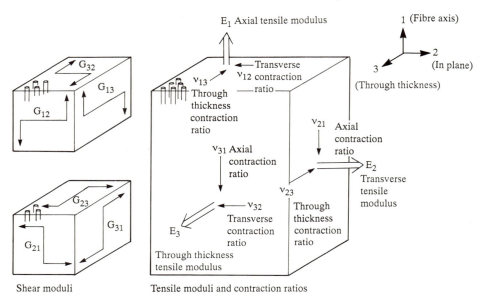

Figure 6.4 Stiffness characteristics of uniaxial laminates

Not all these are independent. In respect of the shear modulus values, geometry requires that $G_{13} = G_{31}$, $G_{12} = G_{21}$ and $G_{23} = G_{32}$. Further, G_{13} is very similar to G_{12}, the difference being associated with the conformation of the resin rich layers (Figure 6.5).

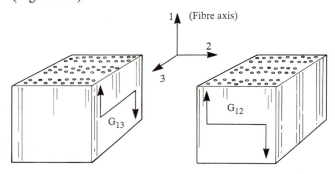

Figure 6.5 Microstructures in shear test. (a) Resin rich interlayer in series. (b) Resin rich interlayer in parallel

On the same basis, as outlined in Appendix 14, we can estimate the shear modulus through the thickness, G_{13}, as 90% of the shear modulus in the plane G_{12} (Appendix 15). By similar arguments $E_3 = 0.97 E_2$.

The independent characteristics usually measured are as follows:

E_1 Axial tensile modulus.
E_2 Transverse tensile modulus.
v_{12} Transverse contraction on uniaxial extension.
v_{21} Axial contraction in transverse extension.
v_{23} Thickness contraction on transverse extension.
G_{12} In plane shear modulus.
G_{23} Through thickness shear modulus.

Several studies of these properties have been reported for two separate carbon fibre PEEK composites, based on 52% by volume fibre and 61% by volume fibre, corresponding to APC-1 and APC-2 respectively[11–27]. These results, summarized in Appendix 16, come from five independent laboratories.

Table 6.2 indicates working values of the stiffness properties of carbon fibre reinforced PEEK based on these data.

Table 6.2 Stiffness of CF/PEEK at 23°C, based on 61% by volume high-strength carbon fibre

		Stiffness	
E_1	(GN/m^2)	137	(2)
E_2	(GN/m^2)	9.4	(0.5)
E_3	(GN/m^2)	9.1*	
v_{12}		0.33	(0.01)
v_{13}		0.32	(0.02)
v_{21}		0.04	(0.01)
v_{23}		0.40	(0.10)
v_{31}		0.04*	
v_{32}		0.40	(0.10)
G_{12}	(GN/m^2)	5.1	(0.5)
G_{13}	(GN/m^2)	4.7*	
G_{21}	(GN/m^2)	5.1*	
G_{23}	(GN/m^2)	3.2	
G_{31}	(GN/m^2)	4.6*	
G_{32}	(GN/m^2)	3.2*	

* indicates estimated values.
(95 per cent confidence limits are noted in parentheses.)

The single most important property is the axial extensional modulus. On the basis of the stiffnesses of the components (Chapter 2), the law of mixtures gives the theoretical modulus for 61% by volume high strength carbon fibre in polyetheretherketone as follows:

$$E \text{ Theory (GN/m}^2) = 0.61 \times 227 + 0.39 \times 3.6$$
$$= 140 \text{ GN/m}^2$$

Experimentally the laminate is achieving 98% of the theoretical value. Similar high utilization of the inherent fibre properties have been demonstrated for a wide range of fibres, including ultra high modulus fibres[28]. The high utilization of the fibre confirms their strong collimation in the mouldings and their excellent wetting by the matrix. For this most important of properties, and for the stiffness parameters in general, attention to the quality of preimpregnation is rewarded.

Cervenka[29] has made a particular study of the stiffness of carbon fibre/PEEK laminates cut at an angle to the fibre orientation (Figure 6.6). Those results appear to be consistent with theoretical prediction.

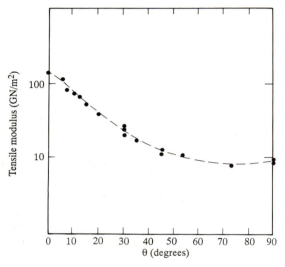

Figure 6.6 Stiffness of uniaxial specimens at an angle θ to the fibre orientation for 61% by volume high strength carbon fibre in PEEK, based on results of Cervenka[30]

6.3.2 Stiffness of multi-angle laminates

With excellent predictive capability established for uniaxial laminates at different angles to the fibre orientation, the properties of multi-angle laminates can confidently be estimated from classical laminate analysis[31].

Several authors provide data for the popular 0/90, ±45 and quasi-isotropic laminates (Appendix 16). Those results are consistent with the predictions based of finite element analysis[32], which predictions take account of the through thickness properties that are ignored in classical laminate theory (Table 6.3).

Cervenka[33,34] has also studied the influence of specimen orientation on the modulus of such laminates. This is dramatic in the case of simple cross plied laminates, where a ±45 laminate tested at 45° to the primary axis effectively becomes a (0,90) laminate. For quasi-isotropic laminates Cervenka finds a maximum of 5% reduction in tensile modulus when testing at 22½° to any fibre axis.

Table 6.3 Comparison of stiffness in ± and quasi-isotropic laminates for carbon fibre/PEEK-based on 61% by volume high strength carbon fibre

Laminate	Prediction of		Tensile modulus GN/m^2
	Laminate analysis	Finite elements[17]	
$(+45, -45)_{NS}$	19.2	18.2	18.7 (0.7)
$(-45, 90, +45, 0)_{NS}$	51.1	48.0	48.4 (1.3)
$(0, 90)_{NS}$	70.3	70.2	74.0 (2.0)

(95% confidence limits in parentheses.)

6.3.3 Creep resistance

When subjected to load for a long time, all materials continue to deform by creep processes. The creep of carbon fibre composites, where a substantial proportion of the fibres are oriented in the direction of the load, is effectively zero[35]. In angle plied, out of plane, transverse or shear loading, where the properties are sensitive to the matrix, significant creep is observed.

Jones[36] provides detailed mechanical property data for PEEK resin in the temperature and time ranges $-80°C$ to $240°C$ and 10^{-4} seconds to 10^8 seconds (1 year). Several studies provide data for creep response in $±45°$ laminates and also in transverse tensile loadings of uniaxial laminates[37–47]. The conclusion from those studies is that the creep performance of composite materials in matrix dominated modes simply mirrors the creep performance of the neat resin, and so can be readily accommodated in design.

Creep resistance is an area where crosslinkable resin systems should have an advantage over linear chain thermoplastic polymers. Surprisingly this is not so. Those who have compared equivalent systems[48–52] observe that, although epoxy systems appear to have an advantage in short term tests, under protracted loading the creep rate with crosslinked materials accelerates and becomes significantly less predictable. This result is attributed to progressive microcracking in the thermoset matrix leading to a gradual reduction of the efficiency with which the inherent reinforcement properties can be translated into the structure. The tough matrices formed from linear chain polymers, and from polyetheretherketone in particular, have a high resistance to such microcracking thereby conferring excellent long term dimensional stability on their composites at normal operating temperatures.

6.4 Dimensional stability and morphological integrity

Crystalline morphology is a factor in reducing the permeation of reagents into the matrix so eliminating any tendency to swelling. Crystallinity also enhances the creep resistance of resins, and therefore of the composites on which they are based.

These factors emphasis a preference for semi-crystalline matrix resins, where long term dimensional stability is critical.

Properties such as thermal expansion and creep resistance are sensitive to microcracking behaviour. Resistance to microcracking is associated with the toughness of the matrix, which is therefore as relevant to those properties of the material where the objective is no change of shape as it is to the ultimate strength and durability.

Close control of, and confidence in, the interface between the matrix and the reinforcement are vital to ensure the full translation of the inherent stiffness of the fibres into form stability of a composite structure. No less significant is the local organization of the fibres in the matrix, both in terms of distribution and orientation. These requirements dictate a preference for preimpregnated uniaxial tape products as the foundation upon which structures with good dimensional stability can be designed with assurance.

References

6–1   S. W. TSAI AND H. T. HAHN, *Introduction to Composite Materials,* Technomic (1980).

6–2   J. T. HARTNESS, 'An Evaluation of Polyetheretherketon Matrix Composites Fabricated from Unidirectional Prepreg Tape', *SAMPE Journal,* **20**, 5, pp. 26–31 (1984).

6–3   F. N. COGSWELL AND M. HOPPRICH, 'Environmental Resistance of Carbon Fibre Reinforced PEEK', *Composites,* **14**, pp. 3–10 (1983).

6–4   *Ibid.*

6–5   E. CORRIGAN, private communication (1989).

6–6   D. B. CURLESS, D. M. CARLIN AND M. S. ARNETT, 'The Effect of Jet Fuel Absorption on Advanced Aerospace Thermoset and Thermoplastic Composites', 35th International SAMPE Symposium, pp. 2127–2141 (1990).

6–7   J. A. BARNES, I. J. SIMMS, G. J. FARROW, G. WOSTENHOLM AND B. YATES, 'Thermal Expansion Behaviour of Thermoplastic Composite Materials', American Society for Composites, 4th Conference in Blacksburg, (1989).

6–8   *Ibid.*

6–9   *Ibid.*

6–10 J. E. ASHTON, J. C. HALPIN AND P. H. PETIT, *Primer on Composite Materials,* Technomic (1969).

6–11 See 6–2.

6–12 D. R. CARLILE, D. C. LEACH, D. R. MOORE AND N. ZAHLAN, 'Mechanical Properties of the Carbon Fibre/PEEK Composite APC-2/AS4 for Structural Applications', ASTM Symposium on Advances in Thermoplastic Matrix Composite Materials, Bal Harbour, Florida (1987).

6–13 S. L. DONALDSON, 'Fracture Toughness Testing of Graphite/Epoxy and Graphite/PEEK Composites', *Composites,* **16**, 2, pp. 103–112 (1985).

6–14 D. P. JONES, D. C. LEACH AND D. R. MOORE, 'Mechanical Properties of Poly Ether Ether Ketone for Engineering Applications', Speciality Polymers Conference, Birmingham (1984).

6–15 S. M. BISHOP, P. T. CURTIS AND G. DOREY, 'A Preliminary Assessment of a Carbon Fibre/PEEK Composite', RAE Technical Report 84061 (1984).

6–16 F. N. COGSWELL, 'Continuous Fibre Reinforced Thermoplastics, in *Mechanical Properties of Reinforced Thermoplastics',* edited by D. W. Clegg and A. A. Collyer, pp. 83–118, Elsevier (1986).

6–17 N. ZAHLAN AND F. J. GUILD, 'Computer Aided Design of Thermoplastic Composite Structures', ICCM VII, Beijing (1989).

6–18 N. M. ZAHLAN, private communication (1990).

6–19 A. CERVENKA, 'PEEK/Carbon Fibre Composites: Experimental Evaluation of Unidirectional Laminates and Theoretical Prediction of their Properties', *Polymer Composites*, **9**, 4, pp. 263–270 (1988).

6–20 P. VAUTEY, 'Cooling Rate Effects on the Mechanical Properties of a Semi-Crystalline Thermoplastic Composite', *SAMPE Quarterly*, **21**, 2, pp. 23–28 (1990).

6–21 P. J. HINE, B. BREW, R. A. DUCKETT AND I. M. WARD, 'The Fracture Behaviour of Carbon Fibre Reinforced Poly (Ether Etherketone)', *Composites Science and Technology*, **33**, 1, pp. 35–72 (1988).

6–22 E. SILVERMAN, R. A. CRIESE AND W. C. FORBES, 'Property Performance of Thermoplastic Composites for Spacecraft Systems', *SAMPE Journal*, **25**, 6, pp. 38–47 (1989).

6–23 D. R. MOORE, I. M. ROBINSON, N. ZAHLAN AND F. J. GUILD, 'Mechanical Properties for Design and Analysis of Carbon Fibre Reinforced PEEK Composite Structures', in *Materials and Processing – Move into the 90's, Proc. 10th International European Chapter Conf of SAMPE*, Materials Science Monographs, **55**, 167–176, Elsevier (1989).

6–24 G. JERONIMIDIS AND T. PARKYN, *Residual Stresses in Thermoplastic Matrix Cross-Ply Laminates*, University of Reading (1987).

6–25 T. NAGUMO, H. NAKAMURA, Y. YOSHIDA AND K. HIRAOKA, 'Evaluation of PEEK Matrix Composites', 32nd International SAMPE Symposium, pp. 486–497 (1987).

6–26 S. GROSSMAN AND M. AMATEAU, 'The Effect of Processing on a Graphite Fibre/ Polyetheretherketone Thermoplastic Composite', 33rd International SAMPE Symposium, pp. 681–692 (1988).

6–27 A. DUTHIE, 'Engineering Substantiation of Fibre Reinforced Thermoplastic Composites for Aerospace Primary Structure', 33rd International SAMPE Symposium, pp. 296–307 (1988).

6–28 J. A. BARNES AND F. N. COGSWELL, 'Thermoplastics for Space', 4th International Congress on Space Materials in Space Environment, 6–9 September 1988, Toulouse, *SAMPE Quarterly*, **20**, 3, pp. 22–27 (1989).

6–29 See 6–19.

6–30 See 6–19.

6–31 N. M. ZAHLAN, private communication (1990).

6–32 See 6–17.

6–33 A. CERVENKA, 'Composite Modelling: Polyetheretherketone/Carbon Fibre Composite (APC-2) Experimental Evaluation of Cross Plied Laminates and Theoretical Prediction of their Properties', Rolduc Polymer Meeting–2, Limburg (1987).

6–34 A. CERVENKA, 'Mechanical Behaviour of Quasi-isotropic PEEK/Carbon Fibre Laminates: Theory/Experiment Correlation', International Conference 'Advancing with Composites' Milan (1988).

6–35 See 6–12.

6–36 See 6–14.

6–37 See 6–12.

6–38 G. R. BELBIN, 'Thermoplastic Structural Composites – A Challenging Opportunity', I.Mech.E. 20th John Player Lecture, London (1984).

6–39 E. M. WOO AND J. C. SEFERIS, *Thermal Sonic Analysis of Polymer Matrices and Composites*, University of Washington (1985).

6–40 X. XIAO, 'Viscoelastic Characterization of Thermoplastic Matrix Composites', Ph.D Thesis, Free University of Brussels (1987).

6–41 A. HOROSCHENKOFF, J. BRANDT, J. WARNECKE AND O. S. BRULLER, 'Creep Behaviour of Carbon Fibre Reinforced Polyetheretherketone and Epoxy Resin', SAMPE Conference, Milan, 'New Generation Materials and Processes' pp. 339–349 (1988).

6–42 X. XIAO AND A. H. CARDON, 'Influence of Phase Interaction on the Properties of APC-2 in Comparison with the Pure PEEK Resin', Free University of Brussels (1988).

6–43 C. HIEL, 'Creep and Creep Recovery of a Thermoplastic Resin and Composite', *Proc. Am. Soc. for Composites*, 3rd Technical Conf., pp. 558–563, Technomic (1988).

6–44 A. CARDON, H. F. BRINSON, C. C. HIEL, 'Nonlinear Viscoelasticity Applied for the Study of Durability of Polymer Matrix Composites', ECCM, **3**, pp. 545–550, Elsevier (1989).

6–45 D. H. NGUYEN, S. F. WANG AND A. A. OGALE, 'Compressive and Flexural Creep Deformation in Thermoplastic Composites', 34th International SAMPE Symposium, pp. 1275–1282 (1989).

6–46 X. XIAO, 'Studies of the Viscoelastic Behaviour of a Thermoplastic Resin Composite', *Composites Science and Technology*, **34**, 2, pp. 163–192 (1989).

6–47 R. Y. KIM AND J. T. HARTNESS, 'Time-Dependent Response of AS4/PEEK Composite', SAMPE Technical Conference (1987).

6–48 See 6–38.
6–49 See 6–39.
6–50 See 6–41.
6–51 See 6–47.
6–52 F. N. COGSWELL AND D. C. LEACH, 'Continuous Fibre Reinforced Thermoplastics: a Change in the Rules for Composite Technology', *Plastics and Rubber Processing and Applications*, **4**, pp. 271–276 (1984).

7 The strength of thermoplastic composites

Like dimensional stability, the strength of a composite material is dominated by the properties of the reinforcing fibre. Consequently the strengths of composite materials are anisotropic. As with the elastic modulus, we must consider both shear and extension but, in the case of strength, it is also necessary to distinguish between tension and compression.

7.1 Compression strength

For monolithic materials, compression strength is significantly greater than tensile strength: this is particularly noticeable in brittle substances. In composites, with the exception of the transverse loading mode, compression performance is usually less satisfactory than extension: this trend is especially evident where the reinforcement is highly anisotropic. Inspection of Appendix 2 demonstrates that, in composite materials, the slender fibres are supported from buckling by the less rigid matrix phase. Carbon fibres are themselves strongly anisotropic structures (Chapter 2), and higher modulus and tensile strength are achieved by higher orientation and finer diameter. When placed in compression, there is the possibility of shearing of the graphite plates within the fibres, or of the fibres splitting transversely along lines of weakness in the highly oriented substructure. The slenderness of the high strength fibres makes them more prone to local microbuckling under compression. Within a family of fibres the higher the stiffness and tensile strength of a fibre, the more likely is that fibre to yield a composite with reduced compression performance.

Some structures require only tensile strength, but many must also bear compression loads. In particular a structure that is to resist flexure, such as an aircraft wing, must experience tension on one surface and compression on the other. In any such structure ultimate performance is determined by the weakest response, and in many structures this weakest feature will be compression strength.

While the tensile strength of a well impregnated uniaxial laminate is readily predicted from the basic properties of the reinforcing fibres, the compression strength of composite materials is also sensitive to the properties of the matrix resin and the organization of the reinforcing fibres. There are many ways in which composite materials can fail in compression. Leeser and Leach[1] identify the

kling, fibre shear, longitudinal splitting, delamination,
:ar crippling. One consequence of this variety of failure
of laboratory measurements of compression strength to
:xpression of this is dependence on gauge length: the
higher the strength[2,3]. There can also be differences in
ow the gauge length is supported and how the load is
1% by volume of high strength carbon fibre in PEEK,
ord values varying from 1,100 to 1,400 MN/m² dependent
.od. Where such discrepancies are evident, I prefer to take
most appropriate working value, and this is given by the
IITRI (Illinois Institute of Technology Research Institute) test.

An account of different investigations is given in Appendix 17, from which working values of the compression strength of carbon fibre reinforced polyetheretherketone are deduced in Table 7.1.

Table 7.1 IITRI compression strength of CF/PEEK, based on 61% by volume high strength carbon fibre at 23°C[9-17]

Laminate	Compression strength MN/m²
Uniaxial – along fibres $(0)_N$	$1,100 \pm 100$
Uniaxial – transverse $(90)_N$	220 ± 40
Cross-plied $(0/90)_{NS}$	670 ± 50
Quasi-isotropic $(45, 90, -45, 0)_{NS}$	630 ± 50

The compression strength of simple uniaxial laminates of carbon fibre/PEEK is generally agreed as being about 30% lower than that of equivalent epoxy composites. As Leeser and Leach[18] point out, the most probable cause of this shortfall is the relatively low shear modulus and yield strength of linear polymers in comparison with highly crosslinked systems. However, as we shall note in Section 7.5, and more particularly in Chapter 8, for thermoplastic composites the technologically critical property of open hole compression strength is not compromised, the post buckling compression strength is significantly superior, and post impact compression strength is outstanding.

Compression performance can be improved by using other fibres. The larger diameter, monolithic, glass[19] and alumina[20] fibres give compression strengths for uniaxial composites of 1,800 MN/m² and 1,240 MN/m² respectively.

7.2 The tensile strengths of uniaxial laminates

There are several independent studies of the tensile strength of uniaxial laminates in the direction of, and transverse to, the fibre orientation[21-29]. These results are collected in Appendix 17 and summarized in Table 7.2.

Table 7.2 Axial and transverse tensile strength of VF/PEEK laminates based on 61% by volume high strength carbon fibre at 23°C

Laminate	Tensile stength (MN/m²)
Uniaxial – along fibres $(0)_N$	2,130 ± 100
Uniaxial – transverese $(90)_N$	80 ± 10

In addition, Cervenka[30] provides a detailed study of the tensile strength of uniaxial laminates at different angles to the fibre orientation (Figure 7.1).

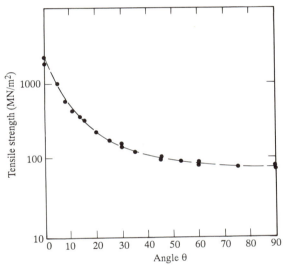

Figure 7.1 Tensile strength of uniaxial laminates at an angle θ to the fibre direction, based on the results of Cervenka, using 61% by volume high strength carbon fibre reinforced PEEK at 23°C[31]

As well as direct tensile measurement, extensive use has been made of flexural testing to describe the apparent tensile modulus and strength of uniaxial laminates. Flexural testing is a complex deformation involving tension, compression and shear, but the results in general agree satisfactorily with simple tensile measurements when the matrix resin has a stiffness above 3.5 GPa, as is the case with PEEK[32]. With lower modulus resins, there is a tendency to premature compression failure of the top surface plies. Flexural testing, and in particular transverse flexure, are extremely valuable as methods of comparative testing of materials or the effects of different environments.

7.3 The tensile strengths of cross-plied laminates

For simple cross-plied laminates in the 0/90 mode[33-35] the tensile strength is, as expected, approximately half that of a uniaxial laminate.

For ±45 laminates simple micromechanics suggests that the tensile strength should be about 120 MN/m² [36]. Experimental determinations[37–42] always give more than twice that value. The stress/strain response for ±45 laminates in tension is markedly non-linear[43] (Figure 7.2).

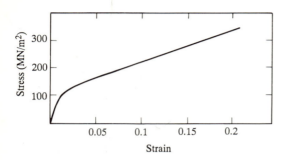

Figure 7.2 Tensile stress/strain response for ±45° laminates of 52% by volume CF in PEEK at 23°C, based on the results of Cogswell and Leach[44]

This non-linearity allows such laminates to be deformed to the order of 20% strain before failure, thereby clearly manifesting the ductility of the matrix phase. Such deformation results in a change of angle from ±45 to ±39 degrees. Cervenka[45] provides a detailed study of the stiffness and strength of angle plied laminates in extension as a function of testing direction. He highlights the limitations of existing theories, whose predictions of strength are dramatically exceeded by preimpregnated carbon fibre/PEEK composites. Although there is little opportunity to exploit this ductility characteristic directly, we may deduce that it is symptomatic of the capability of the resin phase to yield and absorb energy. This ability is particularly significant when a structure is subjected to adventitious out of plane or torsional loadings. Such loadings must be expected in complex assembled structures, where the usual experience is that thermoplastic composites have enhanced strength in comparison to similar structures made from brittle matrix materials.

Cervenka[46] also provides a detailed study of the strengths of quasi-isotropic laminates. In this case experimental measurements exceed the predictions of theory by 20–28%. The larger variation is recorded at an angle of 22.5° to the primary

Table 7.3 Tensile strength of cross-plied and quasi-isotropic laminates

Laminate	Tensile strength MN/m²
(±45)$_S$	300 ± 50
(0/90)$_S$	1,100 ± 50
(−45, 0, +45, 90)$_S$	750 ± 20

fibre orientation where, theoretically, the moulding should have lowest strength. As a result of this the tensile strength is, within experimental error, independent of direction in the laminate.

Various measurements of tensile strength on cross plied and quasi-isotropic laminates are collected in Appendix 17 and summarized in Table 7.3.

7.4 Shear strengths

The measurement of shear strength is more problematic. Most commonly, use is made of a short beam flexural test[47–50] to deduce the apparent interlaminar shear strength. This usually yields values of about $108 \, MN/m^2$, although values as high as $127 \, MN/m^2$ have been reported (Appendix 17). Unfortunately the failure process is not clearly interlaminar shear. An alternative approach is to use off axis tension tests to deduce the shear strength: this provides the low value of 78 to $89 \, MN/m^2$ [51,52]. Duthie[53] reports an in plane shear strength of $175 \, MN/m^2$. The debate about optimum methods of measuring shear strength is not limited to thermoplastic materials but is also apparent in studies of thermosetting materials. Workers at the Royal Aerospace Establishment[54] conclude that the interlaminar shear strength of PEEK composite is significantly greater than that of epoxy composites.

It is reasonable to presume that the interlaminar shear strength will be lower than the intralaminar, or in plane, strength, because of the resin rich interlayer in the former and the slight fibre entanglement in the latter. In the case of angle ply laminates the apparent interlaminar shear strength is reduced by about 25% in comparison with the strength of a uniaxial laminate. This result may be associated with the more defined resin rich interlayer or be an expression of lower strength in the transverse direction. Until more precise tests are performed, we are limited to approximate values for the shear strength of laminates. These are collected in Appendix 17 and summarized in Table 7.4.

Table 7.4 Approximate shear stengths of CF/PEEK laminates based on 61% by volume high strength carbon fibre

	Shear strength MN/m^2
Axial interlaminar shear, σ_{13}	120 ± 10
Axial intralaminar shear, σ_{12}	175 ± 25
Transverse interlaminar shear, σ_{23}	80 ± 10

7.5 Technological tests

All industries have their own technological tests, which usually reflect certain extremes of loading condition. With structural composite materials particular use is made of tests where a hole has been drilled in the sample to represent a fixture. Typical results for open hole tension and compression are shown in Table 7.5.

Table 7.5 Open hole* strengths quasi-isotropic laminates of CF/PEEK, based on 61% by volume high strength carbon fibre[55-7]

	Open hole strength MN/m²
Tension	400 ± 10
Compression	300 ± 10

* Specimen width 38 mm, with 6.35 mm machined hole.

Notched, or open hole, tests of various descriptions have been used by a number of investigators[58-64]. In respect of compression performance these studies conclude that, despite the lower value of uniaxial compression strength, in thermoplastic composites the open hole compression strength is equivalent to that of epoxy systems. There is, however, no obvious benefit in using tough matrix composites for this particular service requirement: the fibre properties dominate the response.

Tough matrix composites are seen to advantage in tests to measure the post buckling compression strength. PEEK matrix composites are reported to have approximately 30% higher strength than epoxy laminates[65]. It is tempting to suggest that this improvement is a manifestation of the enhanced out of plane properties of tough matrix composites with a strong fibre/resin interface.

No technological test is considered of more importance with composite materials than their tolerance of damage. This arena, in which thermoplastic composites are seen to their best advantage, is considered in the following chapter on durability.

References

7–1 D. LEESER AND D. C. LEACH, 'Compressive Properties of Thermoplastic Matrix Composites', 34th International SAMPE Symposium, pp. 1464–1473 (1989).
7–2 R. J. LEE AND A. S. TREVETT, 'Compression Strength of Aligned Carbon Fibre Reinforced Thermoplastics Laminates', Sixth International Conference on Composite Materials (ICCM-6), **1**, pp. 278–287 (1987).
7–3 R. J. LEE, 'Compression Strength of Aligned Carbon Fibre-Reinforced Thermoplastic Laminates', *Composites*, **18**, pp. 1–12, January (1987).
7–4 See 7–1.
7–5 See 7–2.
7–6 See 7–3.
7–7 D. R. CARLILE, D. C. LEACH, D. R. MOORE AND N. ZAHLAN, 'Mechanical Properties of the Carbon Fibre/PEEK Composite APC-2/AS4 for Structural Applications', ASTM STP 1044, American Society for Testing and Materials (1990).

7–8   T. NAGUMO, H. NAKAMURA, Y. YOSHIDA AND K. HIRAOKA, 'Evaluation of PEEK Matrix Composites' 32nd International SAMPE Symposium, pp. 486–497 (1987).

7–9   See 7–1.

7–10 See 7–2.

7–11 See 7–3.

7–12 See 7–7.

7–13 See 7–8.

7–14 Chapter 4.

7–15 A. J. CERVENKA, private communication (1990).

7–16 A. DUTHIE, 'Engineering Substantiation of Fibre Reinforced Thermoplastic Composites for Aerospace Primary Structure', 33rd International SAMPE Symposium, pp. 296–307 (1988).

7–17 P. VAUTEY, 'Cooling Rate Effects on the Mechanical Properties of a Semi-Crystalline Thermoplastic Composite', SAMPE Quarterly, 21, 1, pp. 23–28 (1990).

7–18 See 7–1.

7–19 See 7–3.

7–20 U. MEASURIA,, 'A New Fibre Reinforced Thermoplastic Composite for Potential Raydome Application : PEEK/Alumina', 3rd International Composites Conference, Liverpool (1988).

7–21 See 7–7.

7–22 See 7–8.

7–23 A. CERVENKA, 'PEEK/Carbon Fibre Composites: Experimental Evaluation of Unidirectional Laminates and Theoretical Prediction of their Properties', Polymer Composites, 9, 4, pp. 263–270 (1988).

7–24 J. T. HARTNESS, 'An Evaluation of Polyetheretherketone Matrix Composites Fabricated from Unidirectional Prepreg Tape', SAMPE Journal, 20, 5, pp. 26–31 (1984).

7–25 F. N. COGSWELL, 'Continuous Fibre Reinforced Thermoplastics', in Mechanical Properties of Reinforced Thermoplastics, edited by D. W. Clegg and A. A. Collyer, pp. 83–118, Elsevier (1986).

7–26 See 7–17.

7–27 S. GROSSMAN AND M. AMATEAU, 'The Effect of Processing in a Graphite Fibre/ Polyetheretherketone Thermoplastic Composite', 33rd International SAMPE Symposium, pp. 681–692 (1988).

7–28 M. NARDIN, E. M. ASLOUN, J. SCHULTZ, J. BRAND AND H. RICHTER, 'Investigations on the Interface and Mechanical Properties of Composites based on Non-Impregnated Carbon Fibre Reinforced Thermoplastic Preforms', Proc. 11th International European Chapter of SAMPE, pp. 281–292 (1990).

7–29 E. SILVERMAN, R. A. CRIESE AND W. C. FORBES, 'Property Performance of Thermoplastic Composites for Spacecraft Systems', SAMPE Journal, 25, 6, pp. 38–47 (1989).

7–30 See 7–23.

7–31 See 7–23.

7–32 See 7–8.

7–33 See 7–7.

7–34 S. M. BISHOP, P. T. CURTIS AND G. DOREY, 'A Preliminary Assessment of a Carbon Fibre/PEEK Composite', RAE Technical Report 84061 (1984).

7–35 A. CERVENKA, 'Composite Modelling: Polyetheretherketone/Carbon Fibre Composite (APC-2) Experimental Evaluation of Cross Plied Laminates and Theoretical Prediction of their Properties', Rolduc Polymer Meeting–2, Limburg (1987).

7–36 Ibid.

7–37 See 7–7.

7–38 See 7–8.

7–39 See 7–24.

7–40 See 7–25.

7–41 See 7–34.

7–42 See 7–35.

7–43 F. N. COGSWELL AND D. C. LEACH, 'Continuous Fibre Reinforced Thermoplastics: A Change in the Rules for Composite Technology', Plastics and Rubber Processing and Applications, 4, 3, pp. 271–276 (1984).

7–44 Ibid.

7–45 See 7–35.

7–46 A. CERVENKA, 'Mechanical Behaviour of Quasi-isotropic PEEK/Carbon Fibre Laminates: Theory/Experiment Correlation', International Conference 'Advancing with Composites', Milan (1988).

7–47 See 7–7.

7–48 See 7–8.

7–49 See 7–25.

7–50 G. KEMPE AND H. KRAUSS, 'Processing and Mechanical Properties of Fiber-Reinforced Polyetheretherketone (PEEK)', 14th Congress of the International Council of the Aeronautical Sciences, Toulouse, France (1984).

7–51 See 7–23.

7–52 See 7–17.

7–53 See 7–16.

7–54 See 7–34.

7–55 See 7–1.

7–56 See 7–7.

7–57 NATIONAL MATERIALS ADVISORY BOARD, *The Place for Thermoplastic Composites in Structural Components,* National Research Council NMAB-434 (1987).

7–58 See 7–1.

7–59 See 7–7.

7–60 See 7–57.

7–61 S. M. BISHOP, 'The Mechanical and Impact Performance of Advanced Carbon-Fibre Reinforced Plastics', *Proc. of 1st European Conference on Composite Materials,* ECCM-1, Bordeaux (1985).

7–62 J. RAMEY, A. PALAZOTTO AND J. WHITNEY, *Comparison of Notch Strength Between Gr/PEEK (APC-1 and APC-2) and Gr/Epoxy Composite Material at Elevated Temperature,* Air Force Wright Aeronautical Laboratory (1985).

7–63 S. C. TAN, 'Tensile and Compressive Notched Strength of PEEK Matrix Composite Laminates', *Proc. American Soc for Composite Materials,* First Technical Conference, Dayton, pp. 368–386 (1986).

7–64 D. R. CARLILE AND D. C. LEACH, 'Damage and Notch Sensitivity of Graphite/PEEK Composite', 15th SAMPE Technical Conference, Cincinatti (1983).

7–65 N. M. ZAHLAN, private communication (1990).

8 Durability

No single property of linear chain thermoplastic composite materials has elicited more excitement in both academic and industrial circles than their toughness in comparison with conventional crosslinked thermosetting materials. The pragmatic concept of toughness is expressed in the science of fracture mechanics. Usually structures are designed to carry loads that are well below the ultimate strength of the composite materials. Under such circumstances it is the adventitious impact of foreign bodies that is most likely to cause failure in the composite. Such failure may be complete breakage of the fibres. More commonly it is microcracking of the matrix phase or delamination between the individual plies. Thermoplastic composites have the ability to absorb energy by dissipative mechanisms within the matrix phase, thereby minimizing such damage. Their high toughness also reduces the propagation of any delamination during normal service, and this is of particular importance when the structure is loaded in compression. During their lifetime most structures are subject to intermittent rather than steady loading patterns: in particular the load may vary from tension to compression. It is therefore necessary also to consider the fatigue performance. Finally, materials may be subjected to abrasive loads so that, for a full appreciation of durability, we must also consider the friction and wear resistance of the materials.

8.1 Fracture mechanics

The science of fracture mechanics was originally formulated to explain the durability of brittle monolithic materials such as glass. Surprisingly such theories can also be applied successfully to composite materials with complex morphology[1,2]. In the case of composite materials their anisotropic nature requires consideration of the fracture mechanics both in terms of the mode of deformation (Figure 8.1) and in respect of the orientation of the fibres.

Practical deformations of structures usually include mixed mode loading patterns, where the weakest response will most likely be cause for failure. Crack opening, Mode I, is the measurement that is most widely carried out in the study of composite materials.

There are a wide range of experimental techniques for carrying out measurements of the fracture mechanics parameters. Each has its champions and

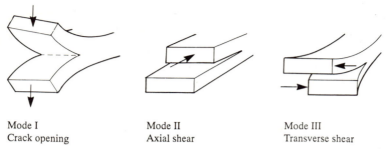

| Mode I | Mode II | Mode III |
| Crack opening | Axial shear | Transverse shear |

Figure 8.1 Cracking modes: Mode I: crack opening; Mode II: axial shear; Mode III: transverse shear

each has its special difficulties[3]. Each also has its particular area of relevance. This text is no place to debate the merits of those different approaches, which is a matter of great consequence to many; instead I shall seek to emphasize what unites them in their interpretation of the behaviour of thermoplastic composites.

8.1.1 Crack propagation in uniaxial laminates

Crick, Leach and Moore[4] have used notched flexure testing to evaluate the fracture toughness of uniaxial laminates as a function of direction of the crack relative to the principal orientation of the fibres in the composite. Although this regime tends to give slightly higher results than more controlled tests, this approach is particularly suitable to comparative tests, since identical sample geometrics can be used for each orientation. Crick and his colleagues identify six principal crack propagation directions (Figure 8.2)[5].

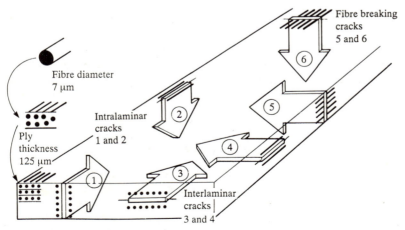

Figure 8.2 Principal directions of crack propagation in a unidirectional composite (adapted from reference 5)

The critical strain energy release rates, G_{IC}, for mode I; crack opening, deformations as a function of crack direction are summarized in Table 8.1.

Table 8.1 Fracture toughness of 61% by volume CF/PEEK at 23°C measured by notched flexure, nominal notch tip radius at initiation 10 μm[6]

Crack direction (Figure 8.2)	Crack mode		G_{IC} kJ/m^2
1	Intralaminar	– axial	5.9
2	Intralaminar	– transverse	5.6
3	Interlaminar	– axial	3.4
4	Interlaminar	– transverse	3.8
5	Fibre breaking	– across plies	406
6	Fibre breaking	– through plies	128

Surprisingly these results show no significant difference in fracture toughness in the axial or transverse directions, in either the intralaminar or interlaminar modes. This implies that the fracture toughness of laminae of different lay up should be qualitatively similar to that of uniaxial plies. Multiangle laminates, however, include internal stresses (Chapter 4), as a result of which the interlaminar fracture toughness of quasi-isotropic laminates is approximately 20% lower than that of uniaxial laminates.

The comparison of interlaminar and intralaminar modes shows the latter to be somewhat tougher. This probably results from slight entanglement of the fibres causing bridging of the crack. The lower toughness in the interlaminar mode should cause cracks to propagate cleanly along the interply regions; however, in such tests, using long double cantilever beam specimens, interlaminar cracks frequently wander into the intralaminar region, causing crack bridging[7].

The fibre breaking cracks absorb considerably more energy than the other modes. Fibre breaking through the plies, where the interlaminar regions are loaded in series, is seen to be less tough than that where the crack is propagated across the plies and the interlaminar regions are loaded in parallel. This again suggests the relative weakness of the resin rich interply region.

8.1.2 Interlaminar crack propagation in different modes

From the study of crack propagation in uniaxial laminates above, the interlaminar crack propagation is seen to be the weakest mode, and so the most likely cause of failure in service. The study of fracture toughness in the crack opening mode, Mode I, and shear mode, Mode II, has attracted a large number of studies[8–35]. These include studies of the early materials, based on 52% by volume of carbon fibre[36–39] and on woven fabric materials[40,41], as well as the standard 61% by volume high strength carbon fibre[42–63]. A variety of different test methods have been used. The results for Modes I and II are collected in Appendix 18.

At this time no direct data are available for the toughness in the transverse shear mode, Mode III. Donaldson, who has measured this property for brittle matrix composites, reported in his verbal presentation of that work[64] that carbon fibre/PEEK was too tough to measure. The highest value that he had been able to measure in other systems was a strain energy release rate of $2\,kJ/m^2$.

Table 8.2 provides working values of the strain energy release rate for carbon fibre reinforced PEEK in the interlaminar modes based on the data available from the literature[65–93].

Table 8.2 Interlaminar fracture toughness of CF/PEEK at 23°C

Mode I Crack Opening G_{IC} kJ/m²	Mode II Axial Shear G_{IIC} kJ/m²	Mode III Transverse Shear G_{IIIC} kJ/m²
2.1 (0.8)	1.8 (0.5)	≫2.0

Standard deviations in parentheses.

While a considerable variation in data is evident in the literature, this is associated with variations in test method and interpretation. Where comparative testing has been carried out, all authors concur that the fracture toughness of CF/PEEK is approximately one order of magnitude greater than that of conventional epoxy resins.

Fractographic studies[94–99] are of considerable assistance in elucidating the toughening mechanisms in composite materials. Such studies emphasize that, in carbon fibre/PEEK composites, the integrity of the interface between matrix and reinforcement is maintained under all loading conditions, and that considerable ductility is evident in the matrix phase, where fibrils of the order of $0.1\,\mu m$ are often evident. Crick[100] associates the local geometry of such ductility with the radiating pattern of the spherulite. He emphasizes that there is no evidence of interspherulitic weakness in carbon fibre/PEEK. The good interface and tough resin also promotes a significant additional toughening mechanism by deflecting crack growth from the resin rich interlayers into the body of the material; this requires breakage of the fibres.

One major variable perceived in fracture mechanics testing is rate: in general the faster the crack propagates, the lower is the toughness. This aspect of the behaviour of carbon fibre reinforced polyetheretherketone has been studied by several authors[101–108]. Typical of such studies are those conducted at the University of Delaware. Smiley and Pipes[109] and Chapman[110], who have studied Modes I and II respectively, find that the toughness of carbon fibre reinforced PEEK starts to reduce at crack displacement rates of about $1\,\mu m/sec$. Maikuma, Gillespie and Wilkins[111] have extended the strain rates at which Mode II fracture toughness can be measured by the use of impact testing. They find significantly less apparent dependence on speed than Chapman. One feature is consistent throughout all the studies: there is similar, or even more dramatic, strain rate dependence with brittle matrix composites. The qualitative improvement achieved with tough thermoplastic matrix composites is maintained irrespective of testing conditions.

8.2 Impact loading

Composite structures are always designed for performance well within their ultimate strength. Failure of such structures is usually associated with adventitious loading, of which impact is the most significant. This field has been intensively investigated[112-126]. In the study of impact performance of composite materials three factors are important: the amount of energy that the material can absorb without breaking, the nature of any internal damage incurred and its effect on residual performance, and the ability to detect any damage on the surface of the structure.

Instrumented studies of impact events[127,128] show force on the impactor rising to a peak and then diminishing with time. From the force/displacement curves the energy absorbed in the event can be determined. The energy to peak force can be associated with the work to start to break reinforcing fibres. The total energy is that absorbed to full penetration.

In respect of the amount of energy which can be absorbed in impact, the response is dominated by the fibre breaking mode. As a first approximation, energy absorbed in impact is proportional to the thickness of the sample to the power 1.5 (Figure 8.3).

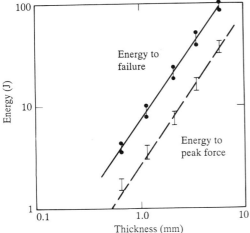

Figure 8.3 Impact energy absorption at 23°C in quasi-isotropic laminates of CF/PEEK composites containing 61% by volume carbon fibre (experimental sample)

Moore and Prediger[129] and Jones and his colleagues[130] have made photographic studies of the impact event that have enabled them to calculate the apparent fracture toughness during impact. These results compare satisfactorily with simple laboratory studies[131,132] of fracture toughness in the fibre breaking mode. It should be noted that the commercial grades of preimpregnated carbon fibre reinforced polyetheretherketone have not been designed to give the highest possible energy absorption in impact. It is possible to obtain higher energy absorption by the use of less tough matrix resins and a less strong matrix/fibre interface: these features encourage crack multiplication and energy dissipation during the generation of free surface within the material. If total energy absorption for a single shot 'shock absorber' is the design criterion, then an alternative system would be more appropriate. The standard commercial grades of preimpregnated CF/PEEK have been designed to maximize energy absorption up to the point where internal damage occurs in the structure.

Internal damage in composite materials is usually evidenced by transverse microcracking in the plies and delamination between plies of different orientation. This usually produces a cone of damage within the sample. The maximum area of such delaminations is conveniently monitored by ultrasonic C-scan (Figure 8.4)[133].

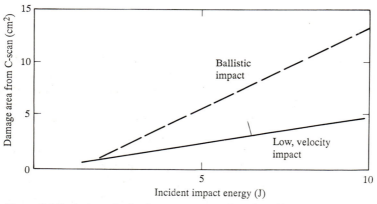

Figure 8.4 Typical results for damage area as a function of impact energy in a quasi-isotropic laminate of CF/PEEK (61% by volume CF), thickness 2 mm, at 23°C (from reference 133)

At the same energy of impact, but at higher velocity of impact, the maximum area of such delamination is increased but the damage tends to be concentrated in the back surface plies (Figure 8.5).

Figure 8.5 Sketch of microcracking and delamination as a function of impact velocity for the same impact energy. (a) Low velocity. (b) High velocity

Reduced internal damage can be achieved by the use of preimpregnated woven fabrics[134]. These also give slightly higher total energy absorption[135]. An alternative approach to reducing delamination in impact is the use of less strong carbon fibres: if a structure is to be subjected to very high velocity, high energy, impacts that will inevitably cause breakage of the fibres, for example micrometeorites, it may be that the function of the structure will be less impaired by using low strength fibres, which allow the impact to pass through the structure without imparting energy that would otherwise cause delamination. The corollory of this conclusion is that the current industry trend to prefer high strength fibres actually places a greater demand upon the resin phase to resist delamination, the consequences of which are considered in Section 8.3.

Finally, it is desirable that any adventitious impact should leave some witness giving warning of damage. With conventional brittle matrix materials, internal delamination can occur without surface damage. With tough thermoplastic matrix composites, impact energies below that critical level, which causes delamination damage within the structure, will cause a dent on the surface that can be detected visually. This surface dent, a local yielding of the matrix phase, is a witness of the ability of the material to dissipate energy within the structure.

8.3 Damage tolerance

For structural composite materials the study of energy absorption in impact is only a beginning. The most important issue to designers is how much strength the structure retains after an adventitious impact event. This tolerance of damage is most important in the case of compression performance. Here any delamination caused by impact may tend to propagate: the compression load will cause thinner sub-laminates to buckle and open any cracks along the loading path (Figure 8.6).

The high value of fracture toughness for carbon fibre/PEEK composites leads to strong resistance, both to initial delamination in an impact event (usually considered a Mode II process) and, more important, exceptional resistance to propagation of such delamination under compression loading (crack opening, Mode I).

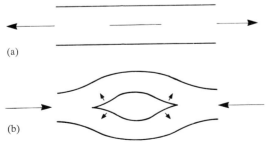

(a)

(b)

Figure 8.6 Influence of subsequent loading on a delamination. (a) Tensile loading has little effect. (b) Compression loading leads to crack opening

The damage tolerance of carbon fibre/PEEK composites has been thoroughly analysed[136-151]. The tests used are technological ones, where results depend on loading path and how the specimen is supported. A widely shared conclusion of these studies is that the fall in residual compressive strength with increasing impact is shallow, and that even severely impacted structures are capable of supporting compression strains of the order 0.6% without catastrophic delamination growth. Further, there is visible damage on the surface of the specimen at lower impact energy than causes any deterioration of properties. These features contrast with typical brittle matrix composites, where the delamination propagates rapidly under compression strains of less than 0.4% and significant internal damage can occur without any surface witness marks being evident. The outstanding damage tolerance of carbon fibre PEEK composites is of particular significance in improving the maintainability of composite structures and thereby reducing their lifetime costs. The laboratory promise of damage tolerance has been confirmed in field trials under arduous circumstances[152].

8.4 Fatigue performance

Besides impact, it is necessary for structures to survive under cyclic loading patterns. Fatigue is an area where, in comparison to the metals that they replace, composite materials have a positive image. Although there are important differences in detail, the fatigue response of carbon fibre/PEEK composites[153-168] is similar to that of widely characterized epoxy materials, and so should be expected to be satisfactory in service.

There is one difference between tough thermoplastic and brittle thermoset matrix composites in fatigue. That difference has two significant effects: in brittle matrix materials there is a tendency to intraply cracking as a stress release mechanism, whereas in thermoplastic composites that energy is dissipated internally within the structures. The internal dissipation in thermoplastic composite materials appears, in accelerated testing, as heat generation. Temperature rises in excess of 150°C have been recorded during tests at five cycles per second[169]. Clearly such excessive heat generation confuses many laboratory studies. To avoid that confusion, Carlile[170] recommends that the test frequency should not exceed one cycle every 2 seconds, which allows the heat generated internally to be lost to the surroundings and thereby holds any temperature rise within convenient experimental bounds. A typical fatigue curve for a ±45 degree laminate is shown in Figure 8.7.

These results indicate a steady reduction in maximum stress towards a fatigue limit of about one half the original static strength. In the case of matrix dominated lay ups such as ±45 the fatigue limit is actually greater than the static strength of a typical epoxy laminate. Fatigue testing at higher frequency[172] suggests that thermoplastic composites are marginally inferior to brittle matrix systems, but this is almost certainly a consequence of internal heating.

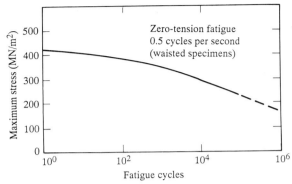

Figure 8.7 Maximum stress versus number of cycles to failure for carbon fibre/PEEK, based on 61% by volume high strength carbon fibres in ±45° laminates (based on reference 171)

A second consequence of the difference in dissipative mechanisms is that, whereas a brittle matrix composite begins to lose stiffness as soon as it starts to microcrack during fatigue, no such reduction in stiffness is seen in tough matrix thermoplastic composites. Kim and Hartness[173] and Picasso[174] suggest that carbon fibre/PEEK composites retain more than 98% of their stiffness out to their full fatigue lifetime: the stiffness of epoxy matrix systems under fatigue rapidly degrades to about 80% of the static value. Microcracking during fatigue of thermoplastic composites usually begins to occur at strain values in excess of 0.6%, double the value of conventional epoxy systems, and the number of such cracks is very much reduced in comparison to brittle matrix systems[175]. The suppression of microcracking in tough matrix composites partially reduces their fatigue resistance in comparison to that of brittle matrix systems. That loss appears to be more than offset by the confidence that the residual properties of the structure, up to its fatigue limit, are not substantially degraded by the fatigue process.

The one area of fatigue behaviour where thermoplastic composites give outstanding performance is the resistance to the growth of interlaminar cracks under cyclic loading[176–179]. In this important area significant advantages, attributable to high matrix fracture toughness, are realized.

In a fatigue damage tolerance comparison of various advanced composite material systems Curtis[180] of RAE Farnborough, who had initially criticised the fatigue performance of thermoplastic systems[181], notes the technical superiority of carbon fibre/PEEK in comparison to standard and toughened thermosetting systems. Since such thermosetting systems are known to be superior to metals in respect of fatigue resistance, the auguries for the use of thermoplastic composites in a fatigue environment appear favourable.

8.5 Wear resistance

The phenomenon of wear is a combination of abrasion, adhesion and fatigue. Abrasion can be thought of as impact on a microscale. Tribology is of importance in a wide range of applications. Typical examples are the retention of thread strength in a machined part, abrasion of artificial joints in a biomedical prosthesis, and wear in reciprocating machinery. Wear resistance is important both in order to maintain the serviceability of the material and also in order to minimize wear debris, which might otherwise be a contaminant. Polyetheretherketone resin itself has excellent wear resistance (Chapter 2) and that translates into good wear resistance in composite forms where the fibres are well wetted by, and strongly interfaced with, the matrix. This field has been studied by workers at the Universities of Hamburg-Harburg and Delaware[182-190].

One contribution to wear behaviour is friction. Mody[191] reports that the coefficient of friction for uniaxial laminates in the temperature range up to the glass transition temperature is about 0.3. Above that temperature the coefficient of friction rises to about 0.6 at 250°C. Voss[192] observes a plateau in friction coefficient of about 0.6 above 150°C. The coefficient of friction of CF/PEEK composites based on preimpregnated woven fabrics is significantly lower than that of uniaxial tape composites below the glass transition temperature but rises to a similar level at high temperature[193].

Because we are concerned with anisotropic materials, it is necessary to consider wear behaviour as a function of fibre orientation: parallel, transverse and normal to the axis of the fibre. It is also necessary to distinguish between sliding wear[194-197] against a smooth metal surface, and abrasive wear[198-202] against a deliberately rough surface. Typical values of specific wear rates observed for sliding wear conditions by Cirino and Pipes[203] are reported in Table 8.3.

Table 8.3 Specific wear rates for sliding wear of moulding of carbon fibre/PEEK based on uniaxial tape of 61% by volume high strength carbon fibre[204]

Fibre orientation to sliding direction	Specific wear rate mm³/Nm
Parallel	0.2×10^{-6}
Transverse	0.4×10^{-6}

Pin-on-ring test conditions: pressure × velocity = 1.7 MN/m² × m/s.

Mody[205], using a similar test, also observes double the wear rate in the transverse compared to parallel mode. Normal to the fibres Mody records similar wear rates to the parallel direction, but Cirino records more than an order of magnitude higher. This discrepancy has not received expert attention, but it may be significant that Mody used much less severe loading conditions, which may have been less inclined to damage the normal fibres. Mody also observes significantly lower wear rates in woven fabric composites, which observation may be associated with the reduced friction of such materials.

In respect of abrasive wear, the rate of abrasion naturally depends upon the roughness of the abrader. Under such loadings Cirino[206] reports similar wear rates in the normal and parallel modes, with slightly higher wear when the fibres are transverse to the direction of abrasion. The wear rates in abrasive wear are several orders of magnitude greater than those in sliding wear. In abrasive wear the fibres are rapidly exposed and play a major role in determining the durability of the material. Where comparative testing has been carried out, there is a consensus that tough, thermoplastic, composites give slightly superior wear resistance to conventional thermosetting systems. This sets a useful baseline from practical experience.

The one area of wear resistance where polyetheretherketone composites do not give as good a performance as epoxy materials is fretting wear[207] where there is vibrating motion between two surfaces in intimate contact. Under such conditions Jacobs[208] suggests that high modulus resin systems are to be preferred; however, this may also be a manifestation of heat generation, as the thermoplastic absorbs the energy imparted.

The tribological characteristics of composites have yet to be fully described in fundamental terms. In particular there is a lack of scientific discussion of wear under rolling conditions. Rolling is important for applications such as gear wheels. In these circumstances it is significant to note that excellent wear resistance in service has been a key factor in defining the suitability of carbon fibre reinforced polyetheretherketone for use in reciprocating machinery[209].

References

8–1   J. E. GORDON, *The New Science of Strong Materials,* Penguin Books (1968).

8–2   G. A. COOPER AND M. R. PIGGOTT, 'Cracking and Fracture in Composites', Fourth International Conference on Fracture, Waterloo, June 1977, *Fracture,* **1**, pp. 557–575 (1977).

8–3   *Ibid.*

8–4   R. A. CRICK, D. C. LEACH AND D. R. MOORE, 'Interpretation of the Toughness in Aromatic Polymer Composites using a Fracture Mechanics Approach', 31st National SAMPE Symposium, pp. 1677–1689 (1986).

8–5   *Ibid.*

8–6   *Ibid.*

8–7   D. C. LEACH, D. C. CURTIS AND D. R. TAMBLIN, 'The Delamination Behaviour of Carbon Fibre/PEEK Composites', ASTM Symposium on Toughened Composites, Houston (1985).

8–8   See 8–4.

8–9   See 8–7.

8–10   D. R. CARLILE AND D. C. LEACH, 'Damage and Notch Sensitivity of Graphite/PEEK Composite', 15th SAMPE Technical Conference, Cincinatti (1983).

8–11   S. L. DONALDSON, 'On Fracture Toughness Testing of Graphite/Epoxy and Graphite/PEEK Composites', *Composites,* **18**, 1 (1987).

8–12   S. L. DONALDSON, 'Fracture Toughness Testing of Graphite/Epoxy and Graphite/PEEK Composites', *Composites,* **16**, 2, pp. 103–112 (1985).

8–13   N. SELA AND O. ISHAI, 'Interlaminar Fracture Toughness and Toughening of Laminated Composite Materials: A Review', *Composites,* **20**, 5, pp. 423–435 (1989).

8–14   L. BERGLUND AND T. JOHANNESSON, 'Mixed-mode Fracture of PEEK/Carbon Fibre Composites', First European Conference on Composite Materials, Bordeaux (1983).

8–15   D. C. LEACH AND D. R. MOORE, 'Toughness of Aromatic Polymer Composites', *Composites Science and Technology,* **23**, pp. 131–161 (1985).

8–16   D. R. CARLILE, D. C. LEACH, D. R. MOORE AND N. ZAHLAN, 'Mechanical Properties of the Carbon Fibre/PEEK Composite APC-2/AS4 for Structural Applications', ASTM Symposium on Advances in Thermoplastic Matrix Composite Materials, Bal Harbour, Florida (1987).

8–17   See 8–4.

8–18   B. R. TRETHEWEY JR, L. A. CARLSSON, J. W. GILLESPIE JR AND R. B. PIPES, 'Mode II Interlaminar Fracture During Static and Fatigue Loading', CCM 86–26, September (1986).

8–19   C. Y. BARLOW AND A. H. WINDLE, The Measurement of Fracture Energy in Aligned Composites, University of Cambridge, pp. 1–11 (1986).

8–20   B. L. SMITH, 'Delamination in Laminated Composite Materials', 18th International SAMPE Technical Conference, pp. 604–612 (1986).

8–21   A. J. SMILEY AND R. B. PIPES, 'Rate Sensitivity of Internlaminar Fracture Toughness in Composite Materials', Proc. American Soc. for Composites, First Technical Conference, Dayton, pp. 434–449 (1986).

8–22   J. W. GILLESPIE JR, L. A. CARLSSON AND A. J. SMILEY, Rate Dependent Mode I Interlaminar Crack Growth Mechanisms in Graphite/Epoxy and Graphite/PEEK', University of Delaware (1987).

8–23   S. HASHEMI, A. J. KINLOCH AND J. G. WILLIAMS, 'Interlaminar Fracture of Composite Materials', ICCM-6, 3, pp. 254–264 (1987).

8–24   P. J. HINE, B. BREW, R. A. DUCKETT AND I. M. WARD, 'Failure Mechanisms in Continuous Fibre Reinforced PEEK', ICCM-6, 3, pp. 397–404 (1987).

8–25   P. DAVIES AND F. X. DE CHARENTENAY, 'The Effect of Temperature on the Interlaminar Fracture of Tough Composites', ICCM-6, 3, pp. 284–294 (1987).

8–26   T. J. CHAPMAN, A. J. SMILEY AND R. B. PIPES, 'Rate and Temperature Effects on Mode II Interlaminar Fracture Toughness in Composite Materials', ICCM-6, 3, pp. 295–304 (1987).

8–27   J. W. GILLESPIE JR AND L. A. CARLSSON, 'Buckling and Growth of Delamination in Thermoset and Thermoplastic Composites', CCM Report, pp. 89–28 University of Delaware (1989).

8–28   A. L. GLESSNER, M. T. TAKEMORI, S. K. GIFFORD AND M. A VALLANCE, 'Temperature and Rate Dependence of Delamination in Aerospace Composite Laminates', SPE ANTEC '87, pp. 1454–1457 (1987).

8–29   P. J. HINE, B. BREW, R. A. DUCKETT AND I. M. WARD, 'The Fracture Behaviour of Carbon Fibre Reinforced Poly (Ether Etherketone)', Composites Science and Technology, 33, 1, pp. 35–72 (1988).

8–30   H. MAIKUMA, J. W. GILLESPIE AND J. M. WHITNEY, 'Analysis and Experimental Characterization of the Center Notch Flexural Test Specimen for Mode II Interlaminar Fracture', J. Composite Materials (in press).

8–31   H. MAIKUMA, J. W. GILLESPIE AND D. J. WILKINS, 'Mode II Interlaminar Fracture of the Center Notch Flexural Specimen under Impact Loading', J. Composite Materials (in press).

8–32   L. A. BERGLUND AND T. R. JOHANNESSON, Effects of Deformation Rate on the Mode Interlaminar Crack Propagation in Carbon Fiber/PEEK Composites, Linkoping Institute of Technology, pp. 131–159 (1987).

8–33   P. DAVIES, W. J. CANTWELL, P.-Y. JAR, H. RICHARD, D. J. NEVILLE AND H. H. KAUSCH, 'Cooling Rate Effects in Carbon Fibre/PEEK Composites', 3rd ASTM Symposium on Composite Materials, Orlando (1989).

8–34   J. T. HARTNESS, 'Polyether Etherketone Matrix Composites', 14th National SAMPE Technical Conference, 14, pp. 26–32 (1982).

8–35   G. DOREY, 'Impact Damage in Composites – Development, Consequences and Prevention', ICCM-6, 3, pp. 1–26 (1987).

8–36   See 8–10.
8–37   See 8–11.
8–38   See 8–12.
8–39   See 8–13.
8–40   See 8–34.
8–41   See 8–35.
8–42   See 8–4.
8–43   See 8–7.
8–44   See 8–14.
8–45   See 8–15.
8–46   See 8–16.
8–47   See 8–4.
8–48   See 8–18.
8–49   See 8–19.

8–50   See 8–20.
8–51   See 8–21.
8–52   See 8–22.
8–53   See 8–23.
8–54   See 8–24.
8–55   See 8–25.
8–56   See 8–26.
8–57   See 8–27.
8–58   See 8–28.
8–59   See 8–29.
8–60   See 8–30.
8–61   See 8–31.
8–62   See 8–32.
8–63   See 8–33.
8–64   S. L. DONALDSON, 'Interlaminar Fracture Toughness due to Tearing (Mode III)', Verbal presentation 'APC too Tough to Measure', ICCM-6, **3**, pp. 274–283 (1987).
8–65   See 8–4.
8–66   See 8–7.
8–67   See 8–10.
8–68   See 8–11.
8–69   See 8–12.
8–70   See 8–13.
8–71   See 8–14.
8–72   See 8–15.
8–73   See 8–16.
8–74   See 8–4.
8–75   See 8–18.
8–76   See 8–19.
8–77   See 8–20.
8–78   See 8–21.
8–79   See 8–22.
8–80   See 8–23.
8–81   See 8–24.
8–82   See 8–25.
8–83   See 8–26.
8–84   See 8–27.
8–85   See 8–28.
8–86   See 8–29.
8–87   See 8–30.
8–88   See 8–31.
8–89   See 8–32.
8–90   See 8–33.
8–91   See 8–34.
8–92   See 8–35.
8–93   See 8–64.
8–94   See 8–4.
8–95   R. A. CRICK, D. C. LEACH, P. J. MEAKIN AND D. R. MOORE, 'Fracture and Fracture Morphology of Aromatic Polymer Composites', *J. Materials Science,* **22**, pp. 2094–2104 (1987).
8–96   D. PURSLOW, 'Matrix Fractography of Fibre-Reinforced Thermoplastics; Part I: Peel Failures', *Composites,* **18**, 5, pp. 365–374 (1987).
8–97   D. PURSLOW, 'Matrix Fractography of Fibre-Reinforced Thermoplastics, Part 2. Shear Failures', *Composites,* **19**, 2, pp. 115–126 (1988).
8–98   D. PURSLOW, 'Fractography of Fibre-Reinforced Thermoplastics, Part 3. Tensile Compressive and Flexural Failure', *Composites,* **19**, 5, pp. 358–366 (1988).
8–99   C. F. BUYNAK, T. J. MORAN AND S. DONALDSON, 'Characterization of Impact Damage of Composites', *SAMPE J.,* **24**, 2, pp. 35–39 (1988).
8–100   See 8–95.
8–101   See 8–10.
8–102   See 8–4.
8–103   See 8–26.

8–104 See 8–28.

8–105 See 8–31.

8–106 See 8–32.

8–107 T. J. CHAPMAN, 'Temperature and Rate Effects on Mode II Interlaminar Fracture Toughness in Composite Materials', B.Sc Thesis, University of Delaware (1986).

8–108 S. MALL, M. KATOUZIAN AND G. E. LAW, *Loading Rate Effect on Interlaminar Fracture Toughness of a Thermoplastic Composite,* University of Missouri (1986).

8–109 See 8–21.

8–110 See 8–107.

8–111 See 8–31.

8–112 See 8–10.

8–113 See 8–16.

8–114 A. A. STORI AND E. MAGNUS, 'An Evaluation of the Impact Properties of Carbon Fibre Reinforced Composites with Various Matrix Materials', in *Composite Structures 2,* edited by I. H. Marshall, Paisley College of Technology, pp. 332–348 (1983).

8–115 D. R. MOORE AND R. S. PREDIGER, 'Photographed Impact of Continuous Fibre Composites', Winter Composites Symposium, Seattle (1985).

8–116 D. P. JONES, D. C. LEACH AND D. R. MOORE, 'On the Application of Instrumented Falling Weight Impact Techniques to the Study of Toughness in Thermoplastics', *Plastics and Rubber Processing and Applications,* **6**, 1, pp. 67–69 (1986).

8–117 D. R. MOORE AND R. S. PREDIGER, 'A Study of Low Energy Impact of Continuous Carbon Fibre Reinforced Composites', presented at NRCC/IMRI Symposium Series 'Composites-87', Montreal (1987).

8–118 J. A. NIXON, M. G. PHILLIPS, D. R. MOORE AND R. S. PREDIGER, 'A Study of the Development of Impact Damage in Cross-Ply CF/PEEK Laminates using Acoustic Emission', submitted to *Composite Science and Technology* (1988).

8–119 C. F. BUYNAK, T. J. MORAN AND S. DONALDSON, 'Characterization of Impact Damage of Composites', *SAMPE J.,* **24**, 2, pp. 35–39 (1988).

8–120 P. E. REED AND S. TURNER, 'Flexed Plate Impact Testing, Part 7. Low Energy and Excess Impacts on Carbon Fibre-Reinforced Polymer Composites', *J. Composites,* **19**, 3, pp. 193–203 (1988).

8–121 E. W. GODWIN AND G. A. O. DAVIES, 'Impact Behaviour of Thermoplastic Composites', *Proc. Int. Conf. on Computer Aided Design in Composite Materials,* Southampton, pp. 371–382 (1988).

8–122 J. MORTON AND E. W. GODWIN, 'Impact Response of Tough Carbon Fibre Composites', *Int. J. Composite Structures* (in press).

8–123 D. R. MOORE AND R. S. PREDIGER, 'A Study of Low-Energy Impact of Continuous Carbon-Fiber-Reinforced Composites', *Polymer Composites,* **9**, 5, pp. 330–336 (1988).

8–124 A. D. CURSON, D. C. LEACH AND D. R. MOORE, 'Impact Failure Mechanisms in Carbon Fiber/PEEK Composites', *Journal of Thermoplastic Composite Materials,* **3**, pp. 24–31 (1990).

8–125 P. J. HOGG AND D. LEICY, 'The Impact Properties of Advanced Thermoplastic Composites', *Fibre Reinforced Polymeric Materials,* Bled, Society of Plastics in Rubber Engineers of Yugoslavia (1989).

8–126 A. D. CURSON, D. C. LEACH AND D. R. MOORE, 'Impact Failure Mechanisms in Carbon Fiber/PEEK Composites', *Proc. Am. Soc. for Composites,* 4th Tech. Conference, pp. 410–417, Technomic (1989).

8–127 See 8–115.

8–128 See 8–116.

8–129 See 8–115.

8–130 See 8–116.

8–131 See 8–4.

8–132 See 8–16.

8–133 See 8–7.

8–134 U. MEASURIA AND F. N. COGSWELL, 'Aromatic Polymer Composites: Broadening the Range', *SAMPE Journal,* **21**, 5, pp. 26–31 (1985).

8–135 See 8–4.

8–136 See 8–7.

8–137 See 8–10.

8–138 See 8–10.

8–139 See 8–35.

8–140 S. M. BISHOP, G. D. HOWARD AND C. J. WOOD, 'The Notch Sensitivity and Impact Performance of

(0, ±45) Carbon Fibre Reinforced PEEK', RAE Technical Report 84066 (1984).

8–141 S. M. BISHOP, 'The Notch Sensitivity and Impact Performance of Carbon Fibre Reinforced PEEK', *Proc. of International Symposium on Carbon Fibre Composites,* Erding, West Germany (1984).

8–142 S. M. BISHOP, 'The Mechanical and Impact Performance of Advanced Carbon-Fibre Reinforced Plastics', *Proc. of 1st European Conference on Composite Materials,* ECCM-1, Bordeaux, France (1985).

8–143 T. M. CORDELL AND P. O. SJOBLOM, 'Low Velocity Impact Testing of Composites', *Proc. American Soc. for Composite Materials,* First Technical Conference, Dayton, pp. 297–312 (1986).

8–144 S. LORENTE, 'Damage Tolerance of Composite Shear Panels', 43rd Annual Forum of the American Helicopter Society, St Louis, Missouri (1987).

8–145 C. SPAMER AND N. BRINK, 'Investigation of the Compression Strength After Impact Properties of Carbon/PPS and APC-2 Thermoplastics Materials', 33rd International SAMPE Symposium, pp. 284–295 (1988).

8–146 M. AKAY, 'Fracture Toughness and Post-Impact Residual Capacity of Carbon Fibre Reinforced Composites', *Composites Science and Technology,* **33**, 1, pp. 1–18 (1988).

8–147 M. NAGHI AND G. NEJHAD, *Effects of Open Holes (Molded-In and Drilled) and Low Velocity Impact Damage on the Compressive Strength of Woven Carbon Fiber Reinforced Polymer Composites,* CCM 88–27, University of Delaware (1988).

8–148 P. SJOBLOM AND B. HWANG, 'Compression-After-Impact: The $5,000 Data Point!', 34th International SAMPE Symposium, pp. 1411–1421 (1989).

8–149 NATIONAL MATERIALS ADVISORY BOARD, *The Place for Thermoplastic Composites in Structural Components,* National Research Council NMAB-434 (1987).

8–150 J. C. PRICHARD AND P. J. HOGG, *The Role of Impact Damage in Post-Impact Compression Testing,* Queen Mary and Westfield College, London (1990).

8–151 J. L. FRAZIER AND A. CLEMONS, 'Evaluation of the Thermoplastic Film Interleaf Concept for Improved Damage Tolerance', 35th International SAMPE Symposium, pp. 1620–1627 (1990).

8–152 'THERMOPLASTIC COMPOSITE BELLY SKINS SUCCESSFUL ON C-130', *Composites News,* May/June 1989; *Advanced Composites* (1989).

8–153 See 8–16.

8–154 J. T. HARTNESS AND R. Y. KIM, 'A Comparative Study of Fatigue Behaviour of Poly Ether Ether Ketone and Epoxy Reinforced with Graphite Cloth', 28th National SAMPE Symposium, pp. 55–62 (1983).

8–155 A. C. DUTHIE, *Fibre Reinforced Thermoplastics in Helicopters Primary Structure,* Westland Helicopters (1987).

8–156 R. Y. KIM AND J. T. HARTNESS, 'The Evaluation of Fatigue Behaviour of Polyetheretherketone/ Graphite Composite Fabricated from Prepreg Tape', 29th National SAMPE Symposium (1984).

8–157 B. PICASSO, P. PRIOLO, C. SITZIA AND P. DIANA, 'Stiffness Reduction During Fatigue in a Graphite/Thermoplastic Composite', *Proc. American Soc. for Composite Materials,* First Technical Conference, Dayton, pp. 36–41 (1986).

8–158 T. K. O'BRIEN, 'Fatigue Delamination Behaviour of PEEK Thermoplastic Composite Laminates', *Proc. American Soc. for Composite Materials,* First Technical Conference, Dayton, pp. 404–420 (1986).

8–159 D. C. CURTIS, D. R. MOORE, B. SLATER AND N. ZAHLAN, 'Fatigue Testing of Multi-Angle Laminates of CF/PEEK', 2nd Int. Conf. on Testing, Evaluation and Quality Control of Composites, TEQC 87, **19**, 6, pp. 446–452 (1988).

8–160 P. T. CURTIS, 'An Investigation of the Tensile Fatigue Behaviour of Improved Carbon Fibre Composite Materials', ICCM-6, **4**, pp. 54–64 (1987).

8–161 C. H. BARON AND K. SCHULTZ, 'Fatigue Response of CRFP with Toughened Matrices and Improved Fibres', ICCM-6, **4**, pp. 65–75 (1987).

8–162 B. R. TRETHEWEY, J. W. GILLESPIE AND L. A. CARLSSON, 'Mode II Cyclic Delamination Growth', *J. Composite Materials,* **22**, pp. 459–483 (1988).

8–163 S. ALYAN, J. CHAUDHUR AND J. AVERY, 'Crack Opening and Crack-Growth Resistance of Graphite/PEEK and Graphite/Epoxy Laminates Containing Large Center Cracks', 33rd International SAMPE Symposium, pp. 1090–1100 (1988).

8–164 A. LUSTIGER, 'Interlamellar Fracture and Craze Growth in PEEK Composites under Cyclic Loading', SPE 47th ANTEC, pp. 1493–1495 (1989).

8–165 T. K. O'BRIEN, 'Fatigue Delamination Behaviour of PEEK Thermoplastic Composite Laminates', *Journal of Reinforced Plastics and Composites,* **7**, 4, pp. 341–359 (1988).

8–166 D. R. MOORE, Chapter 10, 'Fatigue of Thermoplastic Composites', pp. 1–57 in *Thermoplastic*

Composite Materials (ed. L. A. Carlsson), Elsevier (1991).

8–167 D. C. CURTIS, M. DAVIES, D. R. MOORE AND B. SLATER, 'The Fatigue Behaviour of Continuous Carbon Fibre Reinforced PEEK', ASTM, Orlando (1989).

8–168 S. KELLAS, J. MORTON AND P. T. CURTIS, *Fatigue Damage Tolerance of Advance Composite Materials Systems'*, Virginia Polytechnic Institute (1990).

8–169 See 8–159.

8–170 See 8–16.

8–171 See 8–159.

8–172 See 8–160.

8–173 See 8–156.

8–174 See 8–157.

8–175 See 8–161.

8–176 See 8–162.

8–177 See 8–163.

8–178 See 8–164.

8–179 See 8–165.

8–180 See 8–168.

8–181 See 8–160.

8–182 M. CIRINO AND R. B. PIPES, 'The Wear Behaviour of Continuous Fibre Polymer Composites', Sixth International Conference on Composite Materials, **5**, pp. 302–310 (1987).

8–183 M. CIRINO, K. FRIEDRICH AND R. B. PIPES, 'Evaluation of Polymer Composites for Sliding and Abrasive Wear Applications', *Composites,* **19**, 5, pp. 383–392 (1988).

8–184 P. B. MODY, T.-W. CHOU AND K. FRIEDRICH, 'Effect of Testing Conditions and Microstructure on the Sliding Wear of Graphite Fibre/PEEK Matrix Composites', *Journal of Materials Science,* **23**, pp. 4319–4405 (1988); *J. of Materials Science,* 23, pp. 4319–4330 (1990).

8–185 M. CIRINO, R. B. PIPES AND K. FRIEDRICH, 'The Abrasive Wear Behaviour of Continuous Fiber Polymer Composites', *J. Material Science,* **22**, pp. 2481–2492 (1987).

8–186 M. CIRINO, R. B. PIPES AND K. FRIEDRICH, *Sliding Wear Behaviour of Polymer Composite,* University of Delaware (1987).

8–187 H. VOSS, *Friction and Wear of PEEK Composites at Elevated Temperatures,* CCM 87–49, University of Delaware (1986).

8–188 P. JACOBS, K. FRIEDRICH, G. MAROM, K. SCHULTE AND H. D. WAGNER, 'Fretting Wear Performance of Glass, Carbon and Aramid Fibre/Epoxy and PEEK Composites', Int. Conf. on Wear Materials, Denver, Colorado (1989).

8–189 P. B. MODY, T.-W. CHOU AND K. FRIEDRICH, 'Abrasive Wear Behaviour of Unidirectional and Woven Graphite Fibre/PEEK Composites', ASTM Symposium (1986).

8–190 M. CIRINO, R. B. PIPES AND K. FRIEDRICH, 'The Effect of Fiber Orientation of the Abrasive Wear Behaviour of Polymer Composite Materials', *Wear,* **121**, pp. 127–141 (1988).

8–191 See 8–184.

8–192 See 8–187.

8–193 See 8–184.

8–194 See 8–182.

8–195 See 8–183.

8–196 See 8–184.

8–197 See 8–186.

8–198 See 8–182.

8–199 See 8–183.

8–200 See 8–185.

8–201 See 8–189.

8–202 See 8–190.

8–203 See 8–182.

8–204 See 8–182.

8–205 See 8–184.

8–206 See 8–185.

8–207 See 8–188.

8–208 See 8–188.

8–209 I. F. GEHRING, 'Carbon Fibre Polymer Composite Materials in Weaving Machinery Engineering – Examples of Practical Applications', *Melliand Texilberichte,* **70**, pp. 413–418 (1989).

9 Temperature sensitivity

In the characterization of mechanical properties the previous three chapters have concentrated on the response of the material at ambient temperature, 23°C. While most structures spend most of their life near such a temperature, it is necessary to consider the response of the material over a range of temperatures that includes low temperatures as well as high. Civilian aircraft materials usually have to perform their duty in the temperature range −50 to +80°C. For high-speed military aircraft skin temperatures up to 170°C may be encountered. High temperatures are also a feature of materials used in engines. There may also be adventitious exposure to very high temperatures in fires, tolerance of which may literally be a question of life or death. At the other end of the temperature scale is the possible use of composites for cryogenic storage of materials, including fuels such as liquid hydrogen. It is therefore necessary to consider the mechanical performance of thermoplastic composites from −270°C to +250°C, which latter temperature is usually considered to be the maximum continuous service temperature for polyetheretherketone in the form of an unfilled resin.

For structural composite materials the glass transition temperature represents a useful cut off point. For practical purposes a maximum working temperature of Tg minus 30°C has been suggested for load bearing structures[1]. For polyetherether-ketone this suggests a design temperature limitation of about 100°C. However, because PEEK is a semi-crystalline resin it is capable of useful service well above that temperature range in lightly loaded applications. The mechanical properties of PEEK resin have been characterized over the temperature range −80°C to +240°C for twelve decades of time[2]. Since it is the resin whose properties are most sensitive to temperature, that characterization provides a vital starting point to the understanding of composite behaviour.

During the last five years there have been a number of programmes designed to develop new composite materials based on thermoplastic matrix resins with higher glass transition temperature, Tg. Typical of such developments are the melt preimpregnated product forms based on specifically designed matrices of the 'Victrex' family. These have included the semi-crystalline system APC (HTX), Tg 205°C, and the amorphous material APC (HTA), Tg 260°C[3-5]. While semi-crystalline systems such as HTX (high temperature crystalline) and ITX (intermediate temperature crystalline) behave in a similar way to PEEK and retain some useful properties above Tg, the amorphous systems, HTA and ITA, which have the advantage of significantly higher glass transition temperature, must not be used above that temperature. Despite the considerable activity in developing such new systems, composites based on PEEK remain the most extensively characterized materials – in the open literature at least.

In the comparison of properties over a wide temperature range it is not always convenient to measure tensile properties directly. There can be difficulties in testing the required long specimens at very low temperatures, and at high temperatures significant problems of gripping the ends of the specimen are encountered. Flexural testing is a more versatile, if less precise, method from which comparative stiffness and strength can be deduced and, in this chapter, data from such testing, as well as from tensile and compression tests, are included. Where both flexure and tensile testing have been used, there is good qualitative agreement as to the influence of temperature on stiffness, but there may be quantitative differences in respect of strength measurements. Because flexure covers both shear and compression as well as extension, the flexural strength of axially aligned composites tends to be lower than the tensile strength, but transverse flexural strength is higher than transverse tension. For detailed results the reader is invited to refer to the publications cited, but, as a guide, the data in this chapter quote apparent properties relative to those at room temperature.

9.1 The effect of temperature on stiffness

Table 9.1 collects typical reports of the stiffness of carbon fibre/PEEK composites relative to their stiffness at 23°C.

The only significant discrepancy in this table is the stiffness of the transverse specimens at high temperature. In his testing Leach[20] used dynamic mechanical analysis and so had a short time scale for his tests. Hartness[21] used tensile tests, and

Table 9.1 Relative stiffness of carbon fibre/PEEK composites as a function of temperature

Fibre orientation	Source (see references)	−268	23	Temperature °C 80	120	Tg	180	240
0, axial	6, 7		1	1.0	0.95		0.9	0.8
	8*		1		0.9		0.9	
	9, 10	1.0	1					
90, transverse	11, 12		1	1.0	0.9		0.4	0.3
	13*		1		0.5		0.2	
	14, 15	1.0	1					
0/90	16, 17		1		1.0			
±45	18*		1		0.7		0.2	
Quasi-isotropic	19*		1		0.95		0.85	

*Based on 52% volume fraction fibre, all other data 61% volume fraction

the stiffness may be presumed to be measured at much longer times. The results of Hartness are also based on 52% volume fraction, whereas Leach used 61%, so that matrix effects should be more significant. The difference between the results is consistent with creep mechanisms which are known to be significant near the Tg of PEEK[22], and have been discussed in Chapter 6. In respect of such creep behaviour, it is important to recognize that Ha[23] reports that this response can be adequately predicted by a knowledge of the matrix behaviour, which has been thoroughly characterized by Jones[24].

Whereas creep is a problem for matrix dominated laminates at high temperature, the time dependent response of fibre dominated lay ups is the same at 120°C as it is at 23°C[25]. Since good composite structure design would naturally arrange to have fibres oriented along the load path, we may conclude that the stiffness of such, correctly designed, laminates will not be affected by temperature in the range −268°C to +80°C. At 120°C there is likely to be a 5% reduction in composite stiffness as the matrix softens. At 180°C, 45°C above the glass transition temperature of the matrix, the stiffness of the composite is reduced by 10–15% and at 240°C, more than 100°C above Tg, the composite still retains 75% of its stiffness at room temperature.

9.2 The influence of temperature on strength

Strength is a more complicated factor: it is necessary to consider compression as well as extension. There are also data available on the apparent short beam shear strength, a matrix dominated property. Typical results are collated in Table 9.2.

Table 9.2 Relative strength of carbon fibre/PEEK composites as a function of temperature

Fibre orientation	Source (see references)	−268	−50	23	80	120	Tg	180	240
0 Tension	26			1		0.85		0.5	
	27*			1		0.8		0.8	
	28, 29	1.6		1					
	30		1.0	1		0.85			
0 Compression	31			1		0.7		0.45	
0 Short beam shear	32			1	0.9	0.85		0.7	0.65
	33		1.2	1	0.8	0.7		0.45	
90 Tension	34			1	0.9	0.8		0.45	
	35*			1		0.55		0.4	
	36, 37	0.95		1					
± 45 Tension	38*			1		0.75		0.65	
	39		0.85	1		0.7			
Quasi-isotropic tension	40*			1		0.95		0.85	

*Based on 52% volume fraction fibre, all other data 61% volume fraction

The results of Hartness[41] are for composite based on 52% by volume of carbon fibre, and all other results are for 61% by volume carbon fibre: this difference may explain the fractionally lower results which he records. The resin sensitive properties – 90° tension, short beam shear, ±45° tension, and uniaxial compression – fall off more rapidly than the fibre dominated properties, especially above the glass transition temperature. There is no significant loss of strength at very low temperature. Indeed there is evidence that the fibre dominated strength properties are significantly enhanced at temperatures close to absolute zero. As a first approximation we can say that, in the temperature range −268 to 120°C, the strength of carbon fibre reinforced PEEK composites based on 61% by volume of high strength carbon fibre is always at least 70% of the value measured at 23°C. At 180°C, 45°C above the glass transition temperature, this composite retains more than 40% of the strength properties that it has at room temperature.

9.3 Toughness and temperature

Toughness is one property that is expected to deteriorate at low temperature as the resin becomes less ductile. In the case of carbon fibre reinforced polyetheretherketone based on 61% by volume high strength carbon fibre there is very little evidence of such deterioration at low temperature, but there is a marked toughening at high temperature (Table 9.3).

Table 9.3 Relative toughness of carbon fibre/PEEK composites as a function of temperature

Test method	Source (see references)	Temperature °C				
		−50	−30	23	80	120
Fracture toughness						
Mode I, G_{IC}	42		0.8	1		1.7
	43			1	1.4	2.0
	44		0.9	1	1.2	2.0
Mode II, G_{IIC}	45		1.1	1		0.85
	46			1	1.2	1.2
	47	0.95		1		1.05
Through plate impact						
Energy to damage	48, 49	1.0		1	1.0	
Energy to break	50, 51	0.8		1	1.2	

The Mode I, crack opening fracture toughness is more sensitive to temperature than Mode II. Neither the resin dominated fracture behaviour nor the fibre dominated through plate impact results show any significant drop in toughness at low temperature. This result is surprising when taken in the context that the ductile/brittle transition temperature for PEEK resin is considered to be about −15°C[52]. Not only are PEEK composites tough, they are tough at temperatures where the polymer is brittle! One explanation that has been offered for such behaviour points out that, in a preimpregnated composite material, the scale on which the resin is distributed is of the order 1 to 10 μm and that very thin layers of materials respond in a more ductile manner than they do in bulk[53]. On the basis of this argument we should anticipate that satisfactory behaviour at low temperature will be sensitive to the quality of impregnation of the fibres, leading to a preference for preimpregnated product forms if low temperature performance is considered to be critical.

9.4 Fire resistance

The ultimate test of high temperature performance is fire resistance. In common with all organic matrix composites, carbon fibre/PEEK is consumed in a high temperature flame, but this material combination gives off very low levels of smoke and toxic gas[54,55]. In comparison with metal, the through thickness thermal conductivity is low, while the in plane conductivity is moderate: CF/PEEK composites heat up more slowly under an intense local flame; they can dissipate heat sideways in preference to conducting it through the thickness. In addition to this dissipation of heat, PEEK polymer chars readily: the composite transforms itself into a carbon–carbon material in a fire. Laminates based on 61% volume fraction of fibre easily outperform aluminium plates of the same thickness as a fire barrier.

In respect of fire and smoke, PEEK has one of the lowest ratings of all polymers (Chapter 2). This polymer was originally designed to meet the stringent requirements of smoke and toxicity demanded of aircraft interiors.

Fire is the most extreme environment that composites are required to resist. Although it would be unfair to claim that carbon fibre reinforced polyetheretherketone is a useful structural material at 1,000°C, the rules for structural performance change under such conditions. The combination of anisotropic thermal conductivity and high char yield allows a 2 mm sheet of such material to withstand a flame at 1,100°C for nearly 2 hours[56], without producing significant levels of smoke or toxic fumes.

References

9–1   D. C. LEACH, F. N. COGSWELL AND E. NIELD, 'High Temperature Performance of Thermoplastic Aromatic Polymer Composites', 31st National SAMPE Symposium, pp. 434–448 (1986).

9–2   D. P. JONES, D. C. LEACH AND D. R. MOORE, 'Mechanical Properties of Poly Ether Ether Ketone for Engineering Applications', Speciality Polymers Conference, Birmingham (1984).

9–3   See 9–1.

9–4   F. N. COGSWELL, D. C. LEACH, P. T. MCGRAIL, H. M. COLQUHOUN, P. MACKENZIE AND R. M. TURNER, 'Semi-Crystalline Thermoplastic Matrix Composites for Service at 350°F-APC (HTX)', 32nd International SAMPE Symposium, pp. 382–395 (1987).

9–5   P. T. MCGRAIL, P. D. MACKENZIE, H. M. COLQUHOUN AND E. NIELD, 'Development of High Temperature Thermoplastics for use in Carbon Fibre Composites', PRI Polymers for Composites Conference, Paper 6 (1987).

9–6   See 9–1.

9–7   M. DAVIES, D. C. LEACH, D. R. MOORE AND R. M. TURNER, 'Mechanical Performance of Semi-Crystalline Thermoplastic Matrix Composites for Elevated Temperature Service', ICCM-6, **1**, pp. 299–309 (1987).

9–8   J. T. HARTNESS, 'An Evaluation of Polyetheretherketone Matrix Composites Fabricated from Unidirectional Prepreg Tape', *SAMPE Journal*, **20**, 5, pp. 26–31 (1984).

9–9   D. EVANS, J. T. MORGAN, R. J. ROBERTSON AND N ZAHLAN, 'The Physical Properties of Carbon Fibre Reinforced PEEK Composites at Low Temperatures', Joint Cryogenic Engineering Conference and the Seventh International Materials Conference, St Charles, Illinois (1987).

9–10   D. EVANS, J. T. MORGAN, S. J. ROBERTSON AND N. ZAHLAN, 'The Physical Properties of Carbon Fibre Reinforced PEEK Composites at Low Temperatures', *Advances in Cryogenic Engineering*, **34**, pp. 25–33 (1988).

9–11   See 9–1.

9–12   See 9–7.

9–13   See 9–8.

9–14   See 9–9.

9–15   See 9–10.

9–16   See 9–1.

9–17   See 9–7.

9–18   See 9–8.

9–19   See 9–8.

9–20   See 9–1.

9–21   See 9–8.

9–22   S. K. HA, Q. WANG AND F.-K. CHANG, 'Viscoplastic Behaviour of APC-2/PEEK Composites at Elevated Temperatures', 35th International SAMPE Symposium, pp. 285–295 (1990).

9–23   *Ibid.*

9–24   See 9–2.

9–25   See 9–1.

9–26   See 9–1.

9–27   See 9–7.

9–28   See 9–8.

9–29   See 9–9.

9–30   D. R. CARLILE, D. C. LEACH, D. R. MOORE AND N. ZAHLAN, 'Mechanical Properties of the Carbon Fibre/PEEK Composite APC-2/AS4 for Structural Applications', ASTM Symposium on Advances in Thermoplastic Matrix Composite Materials, Bal Harbour, Florida (1987).

9–31   *Ibid.*

9–32   See 9–1.

9–33   G. KEMPE AND H. KRAUSS, 'Processing and Mechanical Properties of Fiber-Reinforced Polyetheretherketone (PEEK)', 14th Congress of the International Council of the Aeronautical Sciences, Toulouse (1984).

9–34   See 9–1.

9–35   See 9–7.

9–36   See 9–8.

9–37   See 9–9.

9–38   See 9–7.

9–39   See 9–30.

9–40   See 9–7.

9–41   See 9–8.

9–42   P. DAVIES AND F. X. DE CHARENTENAY, 'The Effect of Temperature on the Interlaminar Fracture of Tough Composites', ICCM-6, **3**, pp. 284–294 (1987).

9–43   S. HASHEMI, A. J. KINLOCH AND J. G. WILLIAMS, 'Effect of the Matrix on the Interlaminar Fracture Behaviour of Composite Materials', PRI Polymers for Composites Conference, Paper 9 (1987).

9–44   A. L. GLESSNER, M. T. TAKEMORI, S. K. GIFFORD AND M. A. VALLANCE, 'Temperature and Rate Dependence of Delamination in Aerospace Composite Laminates', SPE ANTEC '87, pp. 1454–1457 (1987).

9–45   See 9–42.

9–46   See 9–43.

9–47   T. J. CHAPMAN, A. J. SMILEY AND R. B. PIPES, 'Rate and Temperature Effects on Mode II Interlaminar Fracture Toughness in Composite Materials', ICCM-6, **3**, pp. 295–304 (1987).

9–48   See 9–2.

9–49 See 9–30.
9–50 See 9–2.
9–51 See 9–30.
9–52 See 9–2.
9–53 F. N. COGSWELL, 'Microstructure and Properties of Thermoplastic Aromatic Polymer Composites', 28th National SAMPE Symposium, pp. 528–534 (1983).
9–54 D. C. LEACH, P. J. BRIGGS AND D. R. CARLILE, 'Mechanical and Fire Properties of APC-2', SAMPE Technical Conference, Albequerque (1984).
9–55 D. C. LEACH, P. J. BRIGGS AND D. R. CARLILE, 'Mechanical and Fire Properties of Aromatic Polymer Composites', 3rd Symposium on 'Spacecraft Materials in Space Environment', European Space Agency, Noordvijk (1985).
9–56 See 9–1.

10 Environmental resistance

It is not sufficient for a structural composite to have good mechanical properties. To be of practical value, these properties must be retained in the conditions under which the material must perform its service. Such conditions may include exposure to various hostile environments, ranging from simple thermal ageing, through various fluid reagents, to radiation. Ideally the material properties should be invariant under such attack, in practice some change will usually be expected. This chapter explores the nature and magnitude of such changes as reported for carbon fibre reinforced polyetheretherketone.

The semi-crystalline nature of PEEK resin is of particular importance in the context of providing resistance to the wide range of reagents that may be encountered in the aerospace industry. Amorphous thermoplastic composites can give equivalent mechanical properties with higher glass transition temperature, but they are usually more sensitive to fluid environments. Another significant factor in determining environmental resistance is the integrity of the interface between the fibre and the matrix. Any tendency to debonding at the fibre surface offers a ready channel along which fluids can migrate into the interior of the material. The structural integrity of the melt preimpregnated product form of carbon fibre/PEEK is a significant factor in determining resistance to environmental attack.

10.1 Ageing: time, temperature, air and stress

The performance of many structural materials deteriorates with age. Such ageing can be accelerated by heat, oxygen and stress.

Free volume ageing close to the glass transition temperature is well known to increase the stiffness of polymeric resins and reduce their toughness. After 336 hours at 130°C, Ma[1], recorded a small increase in the axial flexural strength of carbon fibre PEEK composites and a 5% reduction in toughness. Hartness[2], observed a 10% increase in the axial flexural strength of carbon fibre/PEEK composite aged for 1,000 hours in air at 148°C, but he reported a 10% drop in the transverse strength after such a history. Measurements of transverse flexural strength after 1,320 hours at 120°C in air showed no significant change in that property[3], but when the temperature was raised to 220°C, some deterioration was evident after 500 hours. This reduction in properties is believed to be due to oxidation. Even that deterioration still allowed carbon fibre/PEEK to retain a transverse flexural strength in excess of $100\,MN/m^2$, a value which few other composite materials aspire to, even when freshly made.

Weight loss is a useful qualitative guide to thermal stability. Prime and Seferis[4] record a 1% weight loss after 1 hour at 400°C in air, but no weight loss under a nitrogen blanket. They further recorded 'only minimal weight loss at temperatures up to 500°C' confirming the studies of Ma[5], who used thermogravimetric analysis at a heating rate of 10°C/min and observed weight loss beginning at 510°C. In the presence of air, Ma observed rapid decomposition at 650°C, but, under a nitrogen blanket, the weight loss of carbon fibre/PEEK composite was limited to 20% at 750°C. Clearly oxygen significantly accelerates ageing in PEEK based composites at high temperature, and Prime and Seferis[6] also note that copper can catalyse such degradation. These results are, of course, exaggerated exposures in comparison to the expected service conditions for the composite material. In conjunction with the experience on samples aged near the glass transition temperature, these studies indicate that PEEK based composites are strongly resistant to ageing.

Stress can also accelerate ageing. Horn[7] exposed samples of composite to 20% of their failure stress for sixty days without causing any degradation in residual strength. Randles[8] went to the extreme of subjecting notched transverse tensile specimens to 65% of their ultimate tensile strength for seven days. He found no deterioration of residual strength. This latter result emphasizes the tenacity of the interface between the matrix and the fibre.

Stresses are not only caused by loads: thermal stress can also be a significant factor. Sykes[9], Funk[10] and Barnes[11] have exposed carbon fibre/PEEK composite to thermal shock, typically in the temperature regime +100°C to −150°C. There was no detectable effect on residual properties, although after 500 cycles some microcracking is observed in such thermoplastic composites. At approximately one microcrack per centimetre per ply this microcracking density is about one tenth of the severity that is observed in thermosetting composites[12]; this result is independently confirmed by Papadopoulos[13]. This modest tendency to microcracking appears to be more associated with temperature excursions above ambient than below. Barnes[14] indicates that he has been able to thermally shock cross-plied carbon fibre/PEEK from room temperature into liquid helium up to 1,000 times without inducing microcracking.

10.2 Permeability of gases

One aspect of environmental resistance is resistance to permeation. Permeability is a function of structural integrity. The smaller the molecule seeking access, the greater is the severity of the test. Evans[15] has studied the permeability of carbon fibre/PEEK composite to hydrogen in the temperature range 25 to 60°C. It takes about 3 days to establish equilibrium permeability of this gas through a 2 mm thick sheet of composite (Table 10.1).

Table 10.1 Permeability of CF/PEEK composite to hydrogen (2 mm thick quasi-isotropic sheet)

Temperature °C	Permeability mol/m s bar
25	7.2×10^{-12}
40	9.4×10^{-12}
60	15.1×10^{-12}

Evans[16] notes that this value was unaffected by thermal cycling, confirming the strong resistance of these materials to microcracking. He concluded that the results were as good as could be expected for a well bonded composite. It is possible to conclude that any diffusion process that occurs in a well made thermoplastic composite will be at the molecular scale rather than exploiting major faults in the material.

10.3 Water resistance

Of all the fluid environments, water is the one that is least easy to avoid. The plasticization of epoxy resins by water, and the consequent suppression of the glass transition temperature in such systems, is a major concern in the use of thermosetting composites. The tendency for water, and more particularly salt water, to play a role in the corrosion of metals is a further well understood concern with structural materials. The properties of polyetheretherketone and its carbon fibre composites demonstrate excellent resistance to degradation due to exposure to water.

Water diffusion into carbon fibre/PEEK composites is slow but Fickian. Wang and Springer[17] model the water pick up of composite material based on measurements on the neat resin. These predictions give good agreement indicating that there is no significant effect on the interface in the composite. Wang and Springer deduce an equilibrium water content of about 0.2% by weight, in which view they are supported by numerous other authors[18-23]. The highest value for simple immersion is reported by Randles[24] on quasi-isotropic samples immersed in water for seventeen months, giving a value of 0.34%. That value was further increased by Pritchard[25], who interspersed saturation at 72°C with thermal spiking to 135°C. Pritchard noted that such thermal spiking was inclined to greatly exaggerated water uptake in epoxy composites owing to matrix damage and debonding at the interface. Forty-four such thermal spikes raised the equilibrium water content of carbon fibre/PEEK to about 0.4%. The water uptake of carbon fibre/PEEK is approximately a fifth that observed with conventional epoxy resin composite systems.

This very modest water uptake causes no obvious effect on properties. Pritchard[26], with the highest water content, reported a 10% reduction in the tensile strength of ±45 laminates. Hartness[27] observed a similar reduction in transverse flexural strength. Such differences are close to the experimental accuracy of the tests. Other workers[28–32] report no significant change in a range of properties including open hole tension and compression[33], fracture toughness[34,35] and standard flexural properties, including testing at 100°C[36].

The effect of water immersion and stress in combination was studied by Horn[37] and Randles[38]. The latter exposed notched transverse tensile specimens to water at 65% of their ultimate tensile strength for 1 week without affecting that interface sensitive property. Horn[39] loaded his open hole quasi-isotropic samples to 20% of their ultimate strength for 60 days with no effect on subsequent open-hole compression or tensile strengths.

Temperature fluctuations in combination with water exposure can also cause deterioration in some composites. Exposure to hot, wet conditions leads to moisture pick up; if such a sample is suddenly cooled so that the water freezes, the expansion of the water as it freezes may cause microcracking in the material. Such microcracking could enable further water to be picked up in the next hot, wet cycle and so forth. Hopprich[40] reports that 3,000 cycles from 80°C/100% relative humidity to −20°C caused no obvious increase in water uptake and had no deleterious effect on residual properties.

As with water, brine appears to have no significant effect on properties, even on stressed samples[41]. Horn[42] observes a slightly higher pick up of brine than distilled water – 0.25% in comparison to 0.18% for equivalent exposures.

Carbon fibre/PEEK composites demonstrate outstanding resistance to water. This is in part due to the strong interface with the fibre. Where that interface has not been so successfully established, in a properly controlled impregnation process, less satisfactory performance may be encountered. Especial care must also be taken with products based on glass fibres, where the interface may be particularly liable to attack. Although immune interfaces have not yet been established for PEEK/glass fibre composites, special protective size systems have been developed, and encouraging performance has been demonstrated for such materials[43].

10.4 Exposure to aviation fluids

Stober and Seferis[44] made a wide ranging study of the fluid absorption characteristics of PEEK polymers and composites. They conclude that fluids with a solubility parameter between 8.8 and 10.3 and a hydrogen bonding index between 2 and 4 have the power to cause swelling in PEEK. That family includes some benzene derivatives and small chlorinated hydrocarbons. Such fluids on their own are highly volatile, so there is little chance of protracted exposure, however, benzene derivatives are found in jet fuel, and small chlorinated hydrocarbons are sometimes important components of paint strippers used in aircraft applications.

From their studies on a range of fluids Stober and Seferis conclude that the absorption characteristics of carbon fibre/PEEK composites are simply scaled from that of the neat resin: there appear to be no affects associated with the fibres or interface region.

10.4.1 Methylene chloride

Methylene chloride is absorbed by amorphous PEEK resin, which crystallizes in its presence[45,46]. Carbon fibre/PEEK composite material also absorbs methylene chloride, dependent on the initial level of crystallinity in the sample. For composites prepared by rapid cooling from the melt to give sub-optimum crystallinity, the weight gain can be as high as 7%[47]. Randles[48] exposed a standard processed quasi-isotropic panel of carbon fibre/PEEK to methylene chloride for 17 months at 23°C and observed 4.5% pick up. Ma[49] working with film-stacked material, reported 4.1% weight gain. Randles[50] observed that exposure to methylene chloride made no significant difference to the axial stiffness and strength of the composite, but that transverse properties were degraded by about 10%. The addition of stress caused further deterioration in the transverse strength: a notched transverse tensile sample loaded to 65% of its ultimate strength failed on contact with methylene chloride[51]; a sample loaded to a more realistic 25% of ultimate strength survived a seven day exposure and retained 75% of its original strength. Although the properties of PEEK composites are degraded by exposure to methylene chloride, standard processed material provides adequate crystallinity to protect the material from gross deterioration. The most obvious sensitivity is in respect of transverse properties and, even after exposure to methylene chloride in the presence of normal working stresses, carbon fibre/PEEK retains a transverse strength that is significantly greater than that of most other composite materials currently in service in their virgin state. The tendency to crystallize in the presence of a swelling agent such as methylene chloride is a useful self protection mechanism.

Methylene chloride is a good solvent for many amorphous thermoplastics, and is sometimes used as a medium for solution impregnation. Environmental concerns are, however, tending to reduce its use.

10.4.2 Paint strippers

The commonest use of reagents such as methylene chloride is in paint strippers. Horn[52] carried out an accelerated lifetime exposure to paint strippers over 60 days, including repeated applications. He observed that carbon fibre/PEEK composite picked up 2.4% by weight during that exposure and retained 90% of its original open-hole tension and compression strength. Cogswell and Hopprich[53] immersed carbon fibre/PEEK in paint stripper for one month and found no change in a range of flexural properties as a result. While paint strippers are, by definition, designed to dissolve organic polymers they have no significant effect on polyetheretherketone matrix composites.

10.4.3 Hydraulic fluids

Hydraulic fluids are piped throughout much of the structure of an aircraft. There is the potential for such pipes to leak and for such leakage to accumulate and remain for protracted periods. Such fluids can corrode many amorphous polymers. The highest weight gains for carbon fibre/PEEK composite in 'Skydrol' hydraulic fluid have been reported by Randles[54], 0.8% after 17 months, and by Stober[55], who achieved 1.1% pick up after exposure of a specially processed, low crystallinity, sample at 70°C. For standard processed material Horn observed a 0.25% weight gain after exposing a quasi-isotropic sample under 20% of its ultimate tensile strength for sixty days. That exposure produced no significant change in open-hole tension and compression strength. Ma[56] concurred with the finding that hydraulic fluids do not degrade the mechanical performance of carbon fibre/PEEK composites.

10.4.4 Jet fuel

Carbon fibre reinforced PEEK composites will absorb about 1% jet fuel[57-61]. This absorption is greatest in cross plied laminates: uniaxial samples are virtually immune to attack[62]. It is assumed that this sensitivity of cross-plied laminates is associated with thermal stresses in the composite (Chapter 4). Curless[63] has ruled out microcracking and wicking along the fibre interface as possible mechanisms of ingress. The highest absorption reported is 1.3% in a quasi-isotropic laminate under a tensile load of 20% of its ultimate tensile strength. After such exposure Horn[64] reported no deterioration in stiffness but a 10% drop in open hole tension and compression strengths. Because jet fuel has many components, Curless[65] used repeated fuel changes in his studies, which achieved 0.85% pick-up at 82°C over ten weeks. This exposure caused no significant change in axial or transverse flexural strength, but there was a 10% deterioration in ±45 tensile strength at 82°C and a 60% reduction in that property at 121°C. Similar results to those of Horn and Curless have been observed by Corrigan[66]. Clearly the high temperature shear properties of a cross plied laminate are significantly affected by the jet fuel, although the properties at normal anticipated service temperature, up to 80°C, suffer only marginal degradation.

Curless[67] used dynamic mechanical analysis to deduce that the glass transition temperature of cross-plied laminated carbon fibre/PEEK is reduced to 106°C in the presence of jet fuel. The Tg of a uniaxial laminate exposed to jet fuel remains virtually unchanged at 137°C. He deduces that the depression of Tg is associated with plasticization of the matrix and, by desorbtion studies, identifies that aliphatic benzene components as the component of jet fuel were preferentially absorbed by the matrix. That result is consistent with the conclusion of Stober and Seferis[68] in respect of their studies on solubility parameter.

Ethylene glycol is a jet fuel additive that corrodes some materials. It has no significant effect upon polyetheretherketone composites[69,70].

Although jet fuel does not plasticize epoxy resin systems, it does appear to attack the matrix/fibre interface in such systems. Curless[71] reports a halving of the transverse strength of epoxy composites exposed to jet fuel. That property is unaffected in PEEK composites. Despite this deficiency, which would make epoxy materials prone to microcracking, thermosetting epoxy composites are known to perform satisfactorily in a jet fuel environment. The superior microcracking resistance of PEEK composites should prove advantageous in service.

Although amorphous thermoplastic composites usually have an inferior environmental resistance to semi-crystalline materials, in respect of jet fuel Curless[72] observes superior retention of high temperature properties for composites based on the high temperature amorphous resin HTA, in comparison with PEEK.

10.5 The space environment

The space environment contains five factors not usually encountered elsewhere: intense radiation, extreme temperature excursions, vacuum, atomic oxygen, and the potential for high velocity impact from micrometeorites. Barnes[73] provides a discussion of these factors, and the behaviour of PEEK composites subjected to them.

The concern about high vacuum is that it may cause volatiles to be evolved from the material. Such volatiles can condense on to delicate instruments and impede their function. Barnes[74] notes that, in testing to standards set by the European Space Agency, carbon fibre reinforced PEEK composites suffer a total mass loss of the order 0.06 per cent: Sykes[75] reports about 0.1% volatiles. These values are approximately one order of magnitude lower than conventional thermosetting composite materials. A major factor in the insensitivity of PEEK composites to vacuum is, of course, the fact that they absorb very little water (Section 10.3 above), a terrestrial environment that it is very hard to avoid before launch. A second factor is the precisely defined chemistry of the material.

Polyetheretherketone and its composites possess outstanding resistance to radiation. Sasuga[76] exposed PEEK composites to electron beam radiation doseage up to 200 MGy with no effect on composite stiffness and strength at normal operating temperatures. Sasuga did observe a slight increase in stiffness of the composite at 180°C (above Tg), and also reported a marginal reduction in toughness at −195°C. Such changes would be consistent with some crosslinking of the matrix.

In practice it is often the combination of environments that poses particular problems. Radiation attack may degrade the matrix. Vacuum may cause such volatiles to be evolved and condense on sensitive instruments. If degraded products were trapped within the matrix, they would alter mechanical performance, potentially plasticizing the composite at high temperature and embrittling it at low temperature. These would be particular concerns if low molecular weight species were evolved. In the case of PEEK, where the tendency is to increase molecular weight, such fears are unfounded.

We have already (Chapter 6 and Section 10.1 above) noted the excellent resistance of PEEK based composites to thermal cycling. In the space environment that thermal cycling ($-160°C$ to $+120°C$ maximum) as a satellite passes from direct sunlight into the shadow of the earth is also combined with radiation. This combination leads to severe microcracking in most composite materials. Sykes[77] and Funk[78], who exposed composites to 10^{10} rads and 500 thermal cycles from -150 to $+100°C$, found that carbon fibre/PEEK reduces this microcracking problem by an order of magnitude. Funk also studied residual properties of a quasi-isotropic laminate after such a simulated space exposure, noting a marginal increase in stiffness of the composite and a 10% reduction in strength.

With respect to atomic oxygen, it has to be accepted that polyetheretherketone is an organic polymer and subject to attack like most other organic polymers. Such attack is a simple etching process. Barnes[79] reports work suggesting that PEEK composite is marginally less sensitive to atomic oxygen attack than epoxy systems, but both materials need protection if they are to undergo prolonged exposure in low earth orbit.

The space environment is not entirely void. There is the natural hazard of micrometeorites and the added, man made, hazard of satellite debris. Although these hazards may be small, collision velocities can be of the order 8km/s. The kinetic energy of such impacts depends upon the square of the velocity. Under such conditions the toughness of PEEK composites is reduced and major delamination is sometimes observed[80]. Despite the tendency for toughness to be reduced, it is reduced from a high initial value.

10.6 Bio-sensitivity

Carbon fibre reinforced polyetheretherketone is being considered as a material system for prosthetic devices. Inflammation of body tissues can be caused by exposure to unsympathetic chemicals, and also by the presence of small particles that the defence mechanisms of the body seek to absorb: large inert bodies are more readily tolerated. The toughness of PEEK composites, and their good wear resistance (Chapter 9), prevent the breaking off of small particles during normal service. The chemical inertness of the resin, including its resistance to water, also leads to the expectation of good behaviour in biological environments.

Williams, McNamara and Turner[81] have reported preliminary studies on the effect of carbon fibre/PEEK composites on living tissue, and the effect of exposure to the biological environment on the composite. Those results are amplified by workers from Howmedica[82] who additionally confirm that repeated steam and radiation sterilization processes have no adverse effect. The biological environment causes no deterioration in the composite. Williams[83] rates the reaction of body tissues to PEEK composites as mild, and comparable to that of conventional implant materials. Carbon fibre/PEEK composites would more properly be described as bio-inert than bio-compatible.

References

10–1 C.-C. M. MA, C.-L. LEE, H.-C. SHEN AND C.-L. ONG, 'Physical Ageing of Carbon Fiber Reinforced Polyether Ether Ketone and Polyphenylene Sulfide Composites (I)', 35th International SAMPE Symposium, pp. 1155–1164 (1990).

10–2 J. T. HARTNESS, 'An Evaluation of Polyetheretherketone Matrix Composites Fabricated from Unidirectional Prepreg Tape', *SAMPE Journal,* **20**, 5, pp. 26–31 (1984).

10–3 F. N. COGSWELL, 'Continuous Fibre Reinforced Thermoplastics' in *Mechanical Properties of Reinforced Thermoplastics,* edited by D. W. Clegg and A. A. Collyer, pp. 83–118, Elsevier (1986).

10–4 R. B. PRIME AND J. C. SEFERIS, 'Thermo-Oxydative Decomposition of Poly (Ether Ether Ketone)', *J. Poly. Sci. C.,* Poly Letters, **24**, p. 641 (1986).

10–5 C. C. M. MA, H.-C. HSIA, W.-L. LIU AND J.-T. HU, 'Studies on Thermogravimetric Properties of Polyphenylene Sulfide and Polyetherther Ketone Resins and Composites', *J. Thermoplastic Composites,* **1**, 1, pp. 39–49 (1988).

10–6 See 10–4.

10–7 W. J. HORN, F. M. SHAIKH AND A. SOEGANTO, *The Degradation of the Mechanical Properties of Advanced Composite Materials Exposed to Aircraft In-Service Environment,* American Institute of Aeronautics and Astronautics Inc., pp. 353–361 (1986).

10–8 G. PRITCHARD AND S. J. RANDLES, 'Long Fibre Thermoplastics for Aerospace Applications', BFF Reinforced Plastics Congress 88, pp. 73–76 (1988).

10–9 G. F. SYKES, J. G. FUNK AND W. S. SLEMP, 'Assessment of Space Environment Induced Microdamage in Toughened Composite Materials', 18th International SAMPE Technical Conference (1986).

10–10 J. G. FUNK AND G. F. SYKES, 'Space Radiation Effects on Poly (aryl-ether-ketone) Thin Films and Composites', *SAMPE Quarterly,* **19**, 3, pp. 19–26 (1988).

10–11 J. A. BARNES AND F. N. COGSWELL, 'Thermoplastics for Space', *SAMPE Quarterly,* **20**, 3, pp. 22–27 (1989).

10–12 See 10–10.

10–13 D. S. PAPADOPOULOS, 'Use of Unbalanced Laminates as a Screening Method for Microcracking', 35th International SAMPE Symposium, pp. 2127–2141 (1990).

10–14 See 10–11.

10–15 D. EVANS, S. J. ROBERTSON, S. WALMSLEY AND J. WILSON, 'Measurement of the Permeability of Carbon Fibre/PEEK Composites', *Cryogenic Materials, 2, Structural Materials,* pp. 755–763 (1988).

10–16 D. EVANS, S. J. ROBERTSON, S. WALMSLEY AND J. WILSON, *Measurement of the Permeability of Carbon Fibre/PEEK Composites,* Rutterford Appleton Laboratory (1989).

10–17 O. WANG AND G. S. SPRINGER, 'Moisture Absorption and Fracture Toughness of PEEK Polymer and Graphite Fiber Reinforced PEEK', *J. Composite Materials,* **23**, 5, pp. 434–447 (1989).

10–18 See 10–2.

10–19 See 10–7.

10–20 F. N. COGSWELL AND M. HOPPRICH, 'Environmental Resistance of Carbon Fibre Reinforced PEEK', *Composites,* **14**, 3, pp. 1–6 (1983).

10–21 J. P. LUCAS AND B. C. ODEGARD, 'Moisture Effects on Mode II Interlaminar Fracture Toughness of a Graphite Fiber Thermoplastic Matrix Composite', ASTM Symposium 'Moisture Effects on Thermoplastic Composites', Bel Harbour (1987).

10–22 E. J. STOBER, J. C. SEFERIS AND J. D. KEENAN, 'Characterization and Exposure of Polyetheretherketone (PEEK) to Fluid Environments', *Polymer,* **25**, pp. 1845–1852 (1984).

10–23 C. C. M. MA AND S. W. YUR, 'Environmental Effect on the Water Absorption and Mechanical Properties of Carbon Fiber Reinforced PPS and PEEK Composites', SPE 47th ANTEC, pp. 1496–1500 (1989).

10–24 See 10–8.

10–25 G. PRITCHARD AND K. STANSFIELD, 'The Thermal Spike Behaviour of Carbon Fibre Reinforced Plastics', ICCM-6, 4, pp. 190–199 (1987).

10–26 *Ibid.*

10–27 See 10–2.

10–28 See 10–7.

10–29 See 10–17.

10–30 See 10–20.

10–31 See 10–21.
10–32 E. J. STOBER AND J. C. SEFERIS, 'Fluid Sorption Characterization of PEEK Matrices and Composites', *Polymer Engineering and Science,* **28**, 9, (1988).
10–33 See 10–7.
10–34 See 10–17.
10–35 See 10–21.
10–36 See 10–20.
10–37 See 10–7.
10–38 See 10–8.
10–39 See 10–7.
10–40 See 10–20.
10–41 See 10–7.
10–42 See 10–7.
10–43 W. P. HOOGSTEDEN, D. R. HARTMAN, 'Durability and Damage Tolerance of S-2 Glass/PEEK Composites', 35th International SAMPE Symposium, pp. 1118–1130 (1990).
10–44 See 10–32.
10–45 See 10–22.
10–46 See 10–32.
10–47 See 10–32.
10–48 See 10–8.
10–49 C.-C. M. MA, C.-L. LEE, C.-L. ONG AND M.-F. SHEU, 'Chemical Resistance of Carbon Fiber Reinforced Polyether Ether Ketone & Polyphenylene Sulfide Composites', 35th International SAMPE Symposium, pp. 1143–1154 (1990).
10–50 See 10–8.
10–51 See 10–8.
10–52 See 10–7.
10–53 See 10–20.
10–54 See 10–8.
10–55 See 10–22.
10–56 See 10–49.
10–57 See 10–7.
10–58 See 10–8.
10–59 See 10–49.
10–60 D. B. CURLESS, D. M. CARLIN AND M. S. ARNETT, 'The Effect of Jet Fuel Absorption on Advanced Aerospace Thermoset & Thermoplastic Composites', 35th International SAMPE Symposium, pp. 332–345 (1990).
10–61 E. CORRIGAN, private communication (1989).
10–62 *Ibid.*
10–63 See 10–60.
10–64 See 10–7.
10–65 See 10–60.
10–66 See 10–61.
10–67 See 10–60.
10–68 See 10–32.
10–69 See 10–7.
10–70 See 10–8.
10–71 See 10–60.
10–72 See 10–60.
10–73 See 10–11.
10–74 See 10–11.
10–75 See 10–9.
10–76 T. SASUGA, T. SEGUCHI, H. SAKAI, T. NAKAKURA AND M. MASUTANI, 'Electron-Beam Irradiation Effects on Mechanical Properties of PEEK/CF Composite', *Journal of Materials Science,* **24**, pp. 1570–1574 (1989).
10–77 See 10–9.
10–78 See 10–10.
10–79 See 10–11.
10–80 J. A. BARNES, private communication (1990).
10–81 D. F. WILLIAMS, A. MCNAMARA AND R. M. TURNER, 'Potential of Polyetheretherketone (PEEK)

and Carbon-Fibre-Reinforced PEEK in Medical Applications', *Journal of Materials Science, Letters,* **6**, pp. 188–190 (1987).

10–82 K. B. KWARTENG AND C. STARK, 'Carbon Fiber Reinforced PEEK (APC-2/AS-4) Composites for Orthopaedic Implants', *SAMPE Quarterly,* **22**, 1, pp. 10–14 (1990).

10–83 See 10–81.

11 The applications of thermoplastic structural composites

Structural materials, to which people must trust their lives, have to undergo a rigorous evaluation before they are put into service. Nowhere is this more true than in the aerospace sector, where the induction time for a new material to be extensively used is at least five, and typically ten, years. In the case of thermoplastic composite materials there is an added problem in respect of the development of appropriate manufacturing technology. The composite manufacturing industry has to make the transition from labour intensive hand lay up and protracted autoclave processes in order to exploit the high rate production potential of these new materials. Chapter 5 demonstrates how the industry has creatively responded to that challenge. In the evaluation stage, thermoplastic structural composites, and in particular preimpregnated carbon fibre reinforced polyetheretherketone, have demonstrated an excellent property profile, reproducibility, and durability: these designed materials are now beginning to take their place in serial production.

Carbon fibre/PEEK was designed as a material for aerospace. Innovative engineers have also demonstrated applications in the fields of prosthetic devices, marine technology, automotive engineering and specialized industrial machinery components.

11.1 Medical uses

One of the most successful operations in modern surgery is the replacement of hip joints. Bone is a living structure and, to be healthy, it needs stress in order to stimulate its growth and regeneration. Artificial hip joints made from stainless steel are stiff in comparison to bone, and this difference can cause the growth stimulating stresses to be reduced in the bone where the bone and prosthetic device join. The result of this loss of stimulus is a gradual deterioration of the bone and the possible need for remedial surgery. With composite materials there is the ability to tailor the anistropy of the fibres in such a manner that the stiffness of the prosthetic device can be designed to meet the needs of the bone[1]: the designer can tailor both the shape and the properties of his device. To be successful in such an application the material must at least be bio-inert, if not bio-compatible. Preliminary studies[2] indicate that carbon fibre/PEEK meets these requirements. The toughness of this material is also of significance, since this allows it to be conveniently machined to shape (Chapter 5) without the danger of creating small fragments that might stimulate unwanted reaction from the body tissues.

Figure 11.1 shows a canine hip joint developed by Pfizer Howmedica. X-ray studies of the device one year after implantation have indicated a satisfactory condition of the surrounding bone[3].

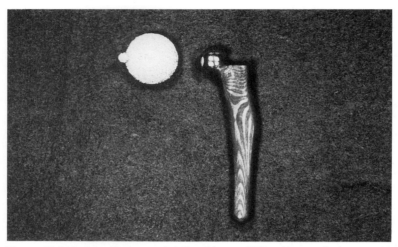

Figure 11.1 Hip joint (Courtesy of J. Dumbleton of Pfizer Howmedica)

As well as their use as implants, thermoplastic composite materials have potential as orthotic devices. De Lollis and Piancastelli[4] note that high stiffness and strength per unit weight can enhance the confidence of the user. With thermoplastic composites there is also the potential for the structure to be conveniently tailored by post forming to meet individual requirements.

11.2 Satellites and launch vehicles

Nowhere are specific properties, stiffness, and strength divided by specific gravity, more highly prized than in space applications: Barnes[5] notes that the cost of carrying a kilogram mass into orbit is of the order £5,000 to £10,000.

Launch vehicles themselves are heavy structures. Typical of the concepts for the next generation of such vehicles is the Horizontal Take Off and Landing vehicle HOTOL, described by Walmsley[6]. The major structure is a fuel tank for liquid hydrogen, which must be capable of withstanding cryogenic shock without microcracking; and it appears (Chapter 10) that CF/PEEK is the only composite material which can do this. Evans[7] confirms that the material does indeed have good resistance to permeation by hydrogen gas. In respect of fabrication such fuel tanks are cylindrical structures of very large size. Cylindrical structures are well suited to filament winding technology. Large filament windings are well suited to thermoplastic technology, because the infinite shelf life of the material means that the winding can be carried out over a long time, and because there is no requirement for large autoclaves in which to cure them. Walmsley[8] considers carbon fibre reinforced PEEK as an appropriate material for such structures.

It is not just mass that contributes to launch costs. Volume is also a significant factor. Space station structures are inevitably bulky items. Spacecraft for deep space exploration are equally bulky and need not have the same structural characteristics as a vehicle designed to penetrate planetary atmospheres. Because thermoplastic composites can be shaped by heat and pressure, there is the potential to transfer the manufacturing process for bulky structures into space[9,10], transporting there, in condensed form, only the materials and equipment necessary to carry out that task. On orbit fabrication should provide an important element of the next phase of mankind's exploration of space, and thermoplastic composites are well suited to play a role in that activity.

Earth launched satellite structures make extensive use of composite materials. Each year the United States Airforce TOPS (ThermOPlastics for Space) programme holds an annual workshop for companies working in this field. Typical of satellite structures being developed is an octagonal box structure, made from carbon fibre reinforced PEEK[11]. That structure (Figure 11.2) is remarkable for

Figure 11.2 Octagonal box structure showing integration of CF/PEEK product forms (Courtesy of R. Garvey)

being built from prepregs based on both conventional high strength carbon fibre and ultra high modulus fibres; it also includes sub-components made from a hybridized staple fibre product form and conventional short fibre injection moulding. Such an approach to optimized structure design emphasizes the potential, optimally, to integrate the diversity of available thermoplastic systems.

Chapters 6 and 10 identified the obvious features of PEEK composite materials making them suitable for satellite structures: dimensional stability, resistance to thermal cycling, resistance to radiation, and the low level of volatile condensable materials. In addition, Silverman and Jones[12] note that carbon fibre reinforced PEEK composites have damping characteristics approximately three times as good as the current qualified grades of epoxy composites for this application. An important series of papers from the TRW Space and Technology Group[13–17], which includes an economic study demonstrating the advantages of thermoplastics[18], concludes that 'future spacecraft structural systems will be able to meet their projected flight requirements by using the superior performance and processing potential of thermoplastic materials over presently existing materials'[19].

11.3 Aircraft structures

Carbon fibre reinforced polyetheretherketone was first developed (1981) at a time of rapidly escalating fuel costs in order to enable commercial aircraft to operate more economically. The outstanding durability of structures made from such materials (Chapter 8) soon attracted military interest, and the United States Airforce has sponsored a series of annual workshops at which the progress of those materials to full scale application has been charted. Typical of the programmes that have been publicly disclosed is the development, by Lockheed, of a generic fighter forward fuselage structure[20] and, by Westlands, of a tailplane for a helicopter[21]. Extensive flight trials of demonstrator components have been in progress. Typical of such demonstrators is a carbon fibre/PEEK inspection panel on a Lynx helicopter that has been in service since 1983 and is showing no deterioration[22]. The importance of inspection panels and hatches is that they are continually being removed from the aircraft so that ground technicians can maintain electronic and hydraulic systems. They are therefore subject to extensive handling and accidental damage. All reports from such trials have been uniformly encouraging.

One of the most significant demonstrator components has been the belly skin of a

Figure 11.3 Landing bay door of F5E (Courtesy of Northrop)

C-130 transport. During take-off and landing there is the possibility of runway debris being kicked up from the wheels, causing impact damage. The carbon fibre/PEEK belly skin of the C-130 is reported to have performed well during extended operation from 'unimproved' Alaskan runways, thereby confirming laboratory assessment of the toughness of the material under low temperature conditions. The underbody durability of carbon fibre/PEEK composites has also been verified by demonstration landing bay doors on military aircraft (Figure 11.3).

Because of the classified nature of military programmes, it is difficult to ascertain the extent to which thermoplastic composites are now being used. However, informed reflection[23] indicates that, besides extensive demonstrator programmes, some aircraft primary structures made from thermoplastic composites have already entered production.

11.4 Marine applications

One of the longest serving demonstration components is on the leading edge of the rudder of a passenger carrying hovercraft. This component, installed by Westlands in 1983, has performed in exemplary fashion[24]. As well as salt spray, normally a highly corrosive environment, there is the problem of impact resistance from sand and stones thrown up by the engines behind which the rudder is mounted. Those engines also provide a fatigue environment.

Besides the natural hazards of the ocean, fire and smoke are the most daunting enemies faced by those 'who go down to the sea in ships'. The fire resistance, and low smoke and toxic gas emission, of carbon fibre reinforced PEEK (Chapter 9), combined with stiffness, strength and durability, have attracted significant attention to this composite as a material system for naval construction. Most notable of such programmes is the vision for building submersibles from thermoplastic composites. As is the case with hydrogen fuel tanks (Section 11.2 above), part of the attraction is to be able to carry out fabrication of very large structures over a long period of time without the need to consider the 'shelf life' of the material and without the requirement to build an autoclave big enough to 'cure' the structure. The ability to weld thermoplastic composite materials may also prove attractive to the shipbuilding industry. The relatively high cost of composite materials such as CF/PEEK in comparison to conventional materials of marine construction may limit their use in that industry initially to specialized applications, but the property spectrum available from these materials may stimulate new design, which, in turn, could provoke the development of lower cost thermoplastic materials designed to the specific needs of the industry.

11.5 Automotive engineering

Automotive applications of materials are notoriously cost sensitive. The value of weight saving in land transportation is significantly less than that in aviation. For successful application it is necessary to seek specific opportunities where performance per unit weight is at a premium. Typical of the applications currently being explored are spring components of cars[25,26] where passenger comfort can be assigned a value justifying the use of high performance composite materials. Another area is Grand Prix racing, where each kilogram of weight saved translates into 1 second in a race. A particular example is the gear selector fork developed by Williams Grand Prix Engineering[27] (Figure 11.4).

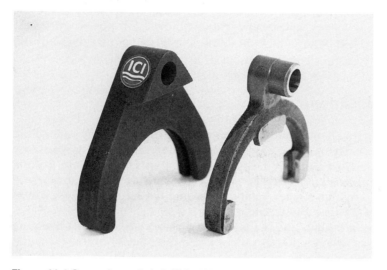

Figure 11.4 Gear selector fork (left) for Williams Formula 1 racing car, together with the forged steel casting that was replaced

This is mass that the driver must move by hand and, in a typical race, he will make 2,500 gear changes. The component must operate in the gearbox, where the oil temperature is about 150°C and local temperature, at the friction surfaces, can rise to 175°C. Williams[28] noted that the gear selector fork, made from preimpregnated carbon fibre reinforced polyetheretherketone, showed a 63% weight saving over the forged steel component that it replaced (aluminium would not perform satisfactorily at these temperatures). The composite component had an improved lifetime expectancy because of reduced wear. Besides these service advantages, the compression moulded and machined component gave a 30% saving in manufacturing costs in comparison to the specialized steel forging it replaced!

11.6 Industrial machinery

The modern age began with the development of industrial machinery in the eighteenth century. At the forefront of that Industrial Revolution was textile technology. The textile machines of today include components of reciprocating mass that operate at up to 20 cycles per second[29]. To increase the speed of reciprocation, and thereby increase the output of the machinery, weight reduction is essential. Gehring[30] describes the development of a weaving machine rapier arm made from carbon fibre reinforced polyetheretherketone (Figure 11.5). Besides the requirement for high stiffness at low weight, carbon fibre/PEEK composite was adopted for this development because it offered the lowest manufacturing costs for high rate production, superior wear resistance, and the ability to recycle offcuts[31]. The rapier arm is a production item.

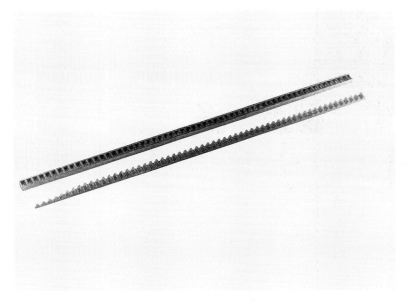

Figure 11.5 Rapier arm developed by Dornier Lindau

Gysin[32] has reviewed the opportunities for advanced thermoplastic matrix composites in mechanical applications, noting a potential market for such materials estimated in excess of £10 million/annum by 1995. As well as the advantages noted by Gehring, Gysin emphasizes the significance of corrosion and fatigue resistance in what will be a dominantly metal replacement market. To illustrate the potential of carbon fibre/PEEK composites, Sulzer has built several prototypes[33]. These include a spring for a weaving machine. The spring has been tested for 87 million cycles without any visible defect or deterioration in performance. Another example developed by Sulzer is a clamp covering for an industrial gas compressor. During its service life this part is loaded by high acceleration forces under elevated temperature in a corrosive environment. In a production trial for this component a rate of one moulding every 3 minutes was attained, but that rate was limited by the availability of cooling. Gysin[34] reported that a manufacturing rate of one part per minute should be achievable. Not only are thermoplastic composites capable of rapid fabrication, but these examples demonstrate how they can be used to develop more effective and more efficient machinery that will reduce energy consumption in manufacturing other goods. Thereby thermoplastics structural composites provide one of the tools to help to revitalize the industrial machinery sector.

References

11–1 K. B. KWARTENG AND C. STARK, 'Carbon Fiber Reinforced PEEK (APC-2/AS4) Composites for Orthopaedic Implants', *SAMPE Quarterly*, **22**, 1, pp. 10–14 (1990).

11–2 D. F. WILLIAMS, A. MCNAMARA (UNIVERSITY OF LIVERPOOL) AND R. M. TURNER, 'Potential of Polyetheretherketone (PEEK) and Carbon-Fibre-Reinforced PEEK in Medical Applications', *Journal of Materials Science Letters*, **6**, pp. 188–190 (1987).

11–3 J. DUMBLETON, private communication (1990).

11–4 A. DE LOLLIS AND L. PIANCASTELLI, *Study and Development of Lower Limb Orthetics for Muscular Distrophy*, Materials Science Monographs, **55**, pp. 51–62 (1989).

11–5 J. A. BARNES AND F. N.COGSWELL, 'Thermoplastics for Space', *SAMPE Quarterly*, **20**, 3, pp. 22–27 (1989).

11–6 S. WALMSLEY, 'The Manufacturing Challenge for Advanced Materials for Hotol', *Proc. PRI Conference 'Automated Composites '88'*, 1/1–1/10 (1988).

11–7 D. EVANS, S. J. ROBERTSON, S. WALMSLEY AND J. WILSON, 'Measurement of Permeability of Carbon Fibre/PEEK Composites', International Cryogenic Conference, China, June (1988).

11–8 See 11–6.

11–9 J. COLTON, J. LYONS, B. LUKASIK, J. MAYER, S. WITTE AND J. MUZZY, 'On-Orbit Fabrication of Space Station Structures', 34th International SAMPE Symposium, pp. 810–816 (1989).

11–10 L. CARRINO, *Composites in Aerospace*, Macplas International, pp. 114–127 (1989).

11–11 R. E. GARVEY, 'Potential for Advanced Thermoplastic Composites in Space Systems', 22nd International SAMPE Technical Conference, pp. 943–956 (1990).

11–12 E. M. SILVERMAN AND R. J. JONES, 'Graphite Thermoplastic Composites for Spacecraft Applications', 33rd International SAMPE Symposium, pp. 1418–1432 (1988).

11–13 *Ibid.*

11–14 E. M. SILVERMAN, R. A. CRIESE AND W. F. WRIGHT, 'Graphite and Kevlar Thermoplastic Composites for Spacecraft Applications', 34th International SAMPE Symposium, pp. 770–779 (1989).

11–15 E. M. SILVERMAN, J. E. SATHOFF AND W. C. FORBES, 'Design of High Stiffness and Low CTE Thermoplastic Composite Spacecraft Structures', *SAMPE Journal*, **25**, 5, pp. 39–46 (1989).

11–16 E. SILVERMAN, R. A. CRIESE AND W. C. FORBES, 'Property Performance of Thermoplastic Composites for Spacecraft Systems', *SAMPE Journal*, **25**, 6, pp. 38–47 (1989).

11–17 E. M. SILVERMAN AND W. C. FORBES, 'Cost Analysis of Thermoplastic Composites Processing for Spacecraft Structures', *SAMPE J.*, **26**, 6, pp. 9–15 (1990).

11–18 *Ibid.*

11–19 See 11–16.

11–20 R. B. OSTROM, S. B. KOCH AND D. L. WIRZ-SAFRANEK, 'Thermoplastic Composite Fighter Forward Fuselage', *SAMPE Quarterly*, **21**, 1, pp. 39–45 (1989).

11–21 A. C. DUTHIE, *Fibre Reinforced Thermoplastics in Helicopter Primary Structure*, Westland Helicopters (1987).

11–22 G. R. GRIFFITHS, private communication (1989).

11–23 M. BURG, *GraFiber News*, 364-3 (1990).

11–24 See 11–22.

11–25 T. AKASAKA, M. MASUTANI, T. NAKAKURA AND H. SAKAI, 'Spring Constants of Eliptic Rings made of Carbon-Fibre-Reinforced Thermoplastics', 33rd International SAMPE Symposium, pp. 670–680 (1988).

11–26 I. KUCH, 'Development of FRP Rear Axle Components', in *Materials and Processing – Move into the 90's, Proc. 20th Int. Eur. Chapter Conf. of SAMPE*, Materials Science Monographs, **55**, pp. 39–50, Elsevier (1989).

11–27 D. WILLIAMS, Williams Grand Prix Engineering, press conference 15 February (1990).
11–28 *Ibid.*
11–29 I. F. GEHRING, 'Carbon Fibre Polymer Composite Materials in Weaving Machinery Engineering – Examples of Practical Applications', *Melliand Textilberichte,* **70**, p. 413 (1989).
11–30 *Ibid.*
11–31 *Ibid.*
11–32 H. J. GYSIN, 'Advanced Thermoplastic Matrix Composites in Mechanical Applications', *Proc. Fibre Reinf. Comp. FRC '90,* I.Mech E, pp. 69–72 (1990).
11–33 *Ibid.*
11–34 *Ibid.*

12 Future developments in thermoplastic structural composites

Thermoplastic structural composites is still an evolving field of materials science. We can look backwards and survey and understand the pathway by which we have arrived at our present position: the purpose of the first eleven chapters has been to chart that understanding. This vantage point is most obviously built upon some five hundred open literature publications concerned with the science and technology of carbon fibre reinforced polyetheretherketone. There are two supports for those publications: the intense industrial activity proving the translation of the promise of these systems into practice, where much of the understanding is retained in-house in order to gain competitive advantage; and the powerful train of academic learning, and teaching, which is making available an increasing number of graduate engineers who will think composites, and indeed think thermoplastic composites, first. We have a secure foundation from which to look forward as well as back.

We can expect to see important developments in the field. These will be associated with new composites, design strategies, processing technology, resource management, and added function. These developments will advance in parallel and with synergy to allow thermoplastic structural composites to play an increasingly important role in engineering.

12.1 New thermoplastic composites

Thermoplastic composite materials embrace a wide field, ranging from the fully impregnated forms based on high loadings of continuous collimated fibres, through random mat stampable sheet products, to short fibre injection moulding compounds. This book considers in detail only one extreme: the production of carbon fibre reinforcement polyetheretherketone at about 60% by volume of reinforcement. It is therefore important to recognize that the materials technology by which that material is made is a generic technology applicable to almost any fibre and resin combination. The concentration in the early days of development on the paradigm, CF/PEEK, has developed science and technology to support a much broader activity. Carbon fibre PEEK composites are fully defined, established materials, which will continue to provide a reference upon which other systems will be able to build.

First of the new systems is solution impregnated amorphous composites resin. To describe solution impregnated materials as new needs a little reflection. Technology for solution impregnation with amorphous thermoplastics, such as the polysulphone family, was described long before the melt impregnation technology that is necessary to handle solvent resistant semi-crystalline polymers. Indeed such materials were evaluated in a wide range of applications and found wanting, usually because of inadequate environmental resistance. Because thermoplastic composites with semi-crystalline matrices can now provide a solution to that problem, users are increasingly willing to consider lower cost amorphous materials for applications where solvent resistance is not so critical. Further, because thermoplastic manufacturing technology has been established to process semi-crystalline composite, and because amorphous materials are actually easier to process in that technology, active consideration is being given to controlling the lifetime service exposure of such materials to hostile environments, and thereby gain the ability to use materials systems which, 10 years ago, were not seriously considered. Solution impregnation of amorphous thermoplastics offers two advantages: price, based on ease of wetting the fibres; and the ability easily to preimpregnate woven broadgoods. As described in Chapter 3, there is a compromise in service properties in comparison with melt preimpregnated product because of the difficulties of eliminating solvent from the product and establishing a fully coherent interface between matrix and fibre. Despite this compromise, the balance of advantages in cost and processability will ensure an important future for this class of material. It is the existence of semi-crystalline systems that liberates the amorphous materials from the constraint of previous prejudice and allows these materials to take their rightful place.

Carbon fibre PEEK composites were designed for applications in the aerospace sector. Chapter 11 shows how these materials are beginning to be exploited as speciality systems in much broader fields. Extending the breadth of application in those fields requires lower cost materials. The technology of melt impregnation is generic: any fibre, any polymer and any level of reinforcement. Thus, today, a new family of thermoplastic composites, 'Plytron', based on typically 40% by volume of glass fibre in resins such as nylon and polypropylene are beginning to become established. These engineering plastics can be used as structural materials in their own right, to add design confidence to compatible stampable sheet materials based on random mat reinforcement, or to provide fusible inserts to be used with injection moulding resins. Although such materials will be of much lower cost than aerospace tailored materials, such as carbon fibre PEEK, they may still appear expensive in comparison to conventional glass mat thermoplastics and moulding compounds. That cost reflects the added value of control of the reinforcement organization and fibre wetting that ensures a high translation of the inherent properties of the composite components. The design confidence that can be established in the performance of structures made from such composites will lead to the adoption of a broad family of thermoplastic composite materials tailored to the needs of different markets.

Conventional stampable sheet materials based on the wetting of bundles of fibres or fibrous mats by melts or particulates are a widely used form of thermoplastic composite. Such materials usually have relatively low fibre content, about 25% by volume. The preceding paragraph noted how design confidence and value could be added to such products by the incorporation of fully wetted, controlled orientation, elements made by processes which ensure a very high level of fibre wetting. Fully preimpregnated thermoplastic composites can further be processed to randomize the fibres and produce a complementary form of stampable sheet product, based on fully wetted fibres and high fibre content. While glass mat thermoplastics based on present technology will continue to be a major, low cost, product form, innovative processing will develop a range of new material forms, based on fully wetted individual fibres at high fibre content that will utilize the established, rapid stamping technologies. These will not be competitive materials to the present family of glass mat thermoplastics, but rather complementary systems that extend the application of this important family.

The field of high rate production is most clearly portrayed by injection moulding. With thermoplastic composites this field is dominated by compounds wherein the fibres are about 0.25 mm long. By fully wetting out fibres with viscous resins it is possible to protect them from attrition during processing, and, as a result, develop long fibre moulding materials that typically retain fibre lengths of 5 mm or more in the moulded article. The technology of impregnating such materials derives from the same materials science origins as the impregnation technology for thermoplastic structural composites such as Aromatic Polymer Composites. Injection moulding compounds, such as the 'Verton' family of long fibre materials, show significantly enhanced durability in comparison to standard injection moulding compounds. These properties are allowing thermoplastic injection moulding materials to be exploited in more demanding applications. Their serviceability, like that of the continuous fibre CF/PEEK prepregs, depends upon the principles of wetting of the fibre surface and interface control inherent in this study.

All these areas are natural points of growth for thermoplastic composite materials. Those areas emphasize a trend towards larger usage of less expensive materials. This will not preclude the further development of low tonnage, high value, materials to meet specialized needs. The vision contains one underlying theme: confidence in performance through control of microstructure.

12.2 Thermoplastic design strategies

Structural composite materials are at present primarily used in the arena of metal replacement. As such, they must work first with design strategies based on metals. Structural design is rightly a conservative field that is evolving from consideration of isotropic materials, through shell like structures that are quasi-isotropic in the plane, to the full use of the anisotropic properties whose exploitation allows composite materials to be used to their greatest advantage. Steady progress continues to be made along this path, and a generation of graduate engineers now leaving university 'think composites' first. In principle, design in thermoplastic materials should be identical to design in thermosetting systems, in which design confidence is limited because of problems with water resistance and brittleness. At this stage designs to use tough, environmentally resistant, systems such as carbon fibre PEEK are not obviously showing the enhanced design confidence in those materials in terms of lower weight, although the potential should be present. Instead the aerospace industry is using similar design for thermoset and thermoplastic systems, and seeking to exploit improved durability by lower maintenance costs in service. As confidence grows we should expect to see new design philosophies emerging.

A particular challenge to the designer will be to consider the out of plane properties of materials in greater detail than heretofore. An embarrassment with some testing of full scale thermoplastic structures has been that the structures have required substantially higher loads to cause failure than was predicted by the design theories based on thermoset experience: this has sometimes led to breaking of expensive test machines. The problem appears to be that the translation of even simple loading patterns into complex structures inevitably produces out of plane loading. It is such adventitious out of plane loading that triggers failure. Thermoplastic composites, such as carbon fibre PEEK, have been designed with a strong interface between resin and reinforcement, in contrast to the weak interface systems on which most practical experience is based. This qualitative improvement in the reproducibility of the composite material, and the consequent high through thickness strength of preimpregnated materials, needs to be incorporated into design practice so that, in future, we shall have a true thermoplastic composite design philosophy.

A particular opportunity for designers will be to conceive of total thermoplastic composite systems. Many components are based on sandwich structures, where skins are laminated to low density areas. Those cores, foams or honeycombs, could potentially be made from the same resin as the composite matrix. This approach should permit the design of more reliable structures, since the interfaces between dissimilar materials, normally a point of weakness, would be eliminated. The innovative designer should also be able to draw on other thermoplastic experience, for example to construct an integral hinge in a component. A structure that is composed of fully compatible material elements would also offer easier lifetime maintainability, and, at the end of its life, would be convenient to recycle.

Design for service must also include the interaction of the material with processing history. At present, this is achieved by an interative process: design the fibre organization, study the change of organization during processing, and redesign the initial lay up. The rules of the interaction between morphology and processing history are becoming established, and it should soon be possible to use model studies to optimize design without trial and error. Ultimately the challenge is to design structure and fabrication process simultaneously.

12.3 Processing technology

Many of the manufacturing processes used to fabricate thermoplastic composite materials have been adapted from existing technology. The two parents have been conventional thermoset composite technology for large structures and metal working for high rate production of simple sections or small parts. The adaptations have been significant and have been justified by the perceived value of the thermoplastic products. In the early years it is necessary for a new material to be processed on existing equipment. As it grows to maturity innovative engineering develops optimized technology to process the new product form. With thermoplastic composites this phase is already evident, with the invention of thermoplastic stretching processes such as 'double diaphragm forming' and assembly technologies such as 'thermoplastic interlayer bonding'. Other distinct thermoplastic strategies are being explored in the area of 'sequential processing', which exploits the potential to reprocess the material a number of times, thereby slowly building up a shape. The rich diversity of processing methods already identified exploit one key feature of thermoplastic materials – the resin chemistry has all been carried out by those people whose business is chemistry. As a result of this feature, the material, at the end of the process, is chemically the same as it was at the beginning. This gives advantages in quality assurance. As the usage of these materials moves from specialized applications to mass production, the stimulus to develop processing technology that most efficiently exploits the thermoplastic character of these materials will increase.

The potential for mass production brings with it opportunities to produce low cost stock shapes. Today composite structures are usually made to a bespoke design. The availability of low cost stock shapes and true thermoplastic assembly technology could provide an alternative way in which composite materials could be applied to solve engineering problems.

In the development of thermoplastic composite technology the integration of product form will have a major role to play. Of the technologies in place today, the use of polymer in film form, as an aid to obtaining thermoplastic bonding of structures, is the most widely employed integration. Sandwich structures, which employ a lower cost short fibre, resin sheet, or foam material as the core, have also been demonstrated. The potential for selective reinforcement of stampable sheet and injection moulding compounds has already been noted (Section 12.1) as a potentially important part of the engineering plastics activity that is just beginning. Successful integration of product form will present a challenge to processors and an opportunity to designers to achieve the most cost effective structure.

Thermoplastic materials inherently possess the potential of reprocessing. With 'difficult' mouldings, this allows the material to be reprocessed to specification if, at the first attempt, incomplete consolidation or some undesired warpage occurs. The potential will also be used to evolve repair strategies that will enhance the lifetime of structural materials.

12.4 Resource management

The performance of a structural material must outlast the usefulness of the structure. It follows then that, when the service lifetime of a structure is expired,

the materials from which it is built still possess value. At that time the ingredients of the structure can potentially be reclaimed. The issue of reclaiming high value materials is also significant in fabrication processes, which inevitably will include some offcuts and scrap. Chapter 5 included a description of how thermoplastic structural materials can be recycled with the addition of extra resin to provide high value moulding compounds. A significant proportion of structures may be reclaimable as stock sheet products by a simple trimming and/or press moulding operation. Such reclaimed material may, because it is already consolidated, prove a preferred material system for the production of very thick stock shapes for subsequent machining: here one would not only be reclaiming the value of the raw material but also some of the value of its previous processing history. With high char yield systems, such as CF/PEEK, it is also possible to consider recycling lifetime expired materials as carbon–carbon composites for higher value applications. The issue of material resource management requires a business planning vision that might include, for example, the concept of material leasing as well as simple sales. Constructing that vision presents a challenge to economists, conservationists and materials scientists alike. Thermoplastic materials systems offer important potential for recycle.

12.5 Added function

Composite materials have other roles to play besides providing structure. Some of those functions, such as heat exchange, are primarily controlled by the fibres. The additional functions, most obviously associated with the matrix, are those of the surface. Surface science is a well developed area for thermoplastic materials. These provide a great many film products that can be described as two surfaces that happen to be joined together with polymer.

The importance of surfaces has already been explored in Chapter 8 when considering the tribological properties of materials and their influence on durability. Another important surface property is cosmetic appearance. One of the well developed arts of thermoplastic materials is the generation of coloured materials: a thin film of such coloured material, thermoplastically bonded on to the surface of the composite, would provide an integral skin as an alternative to painting. Such a strategy of course assumes that there is no wish frequently to recolour the structure. The surface also has a role to play in resistance to environmental attack. Some corrosive agents, such as atomic oxygen, can only be protected against by a shield, which, ideally, should be integrally bonded on to the surface. The art of functionalizing the surface of a material so that it can have desired characteristics is an area with significant potential for innovation.

Structural materials are primarily designed to give passive resistance to mechanical loading. It is possible to incorporate sensors into materials and to monitor their state of deformation at any one time, indeed strain gauges are extensively used during the mechanical property testing of structural materials. As well as sensors, it is possible to incorporate actuators, such as shape memory alloys, into a composite, with a view to changing the state of stress in the structure in response to external stimulus. It is further possible to include piezo-ceramic elements into composite structures, with a view to simultaneously sensing and activating deformation in the structure. Combining sensors and actuators together, with perhaps 1 kilogram of computer for every 10 kilograms of structure, produces a material system that can sense, and react to, a change in its situation – the concept of 'smart' structures. Smart structures are, by definition, composite concepts. Fibres can carry messages as well as loads. The potential for thermoplastic materials in such systems is enhanced by the morphological integrity that results from the use of tough matrix resins and controlled, strong, interfaces. Initially the use of the 'smart' materials concept will simply be to monitor the state of the structure to detect damage. The more reliable the structures, the more reliable will that feedback be. This application will evolve to controlled, dynamic, feedback, for example to damp vibrations in a structure by sensing the vibration, and cancelling it by inducing an out of phase response in the structure. The quality of the feedback information, and so the efficiency of that control, will depend upon an exact knowledge of the response of the structure to load. Finally, the incorporation of discrete actuators within the structure itself, allowing external loading to be balanced by internal stress, calls for the embedding of those actuators within the structure. The insensitivity of thermoplastic composite materials to such discontinuity, a manifestation of the inherent toughness of the matrix, should make them particularly appropriate for such applications. As material science moves towards active materials, thermoplastics will play an increasingly important role.

12.6 Envoy

The processing, properties and performance of thermoplastic Aromatic Polymer Composites derive from the probity of their microstructure. First is the immutability of the resin chemistry, which is defined and controlled by those people whose business is chemistry. The molecular structure of the resin, and in particular its ability to crystallize, determine the range of conditions and environments in which the matrix can give service. That structure also defines the inherent toughness and processability of the system. Next is the predetermined distribution of resin and reinforcement established during preimpregnation. That local distribution is maintained through the processing cycle, where the viscous resin protects the reinforcement fibres from damage. As a result, the finished structure translates the full inherent stiffness and strength of the reinforcement into service potential. Also created during preimpregnation is the vital interface between matrix and reinforcement, whose integrity determines the difference between a composite and a good composite. Through that interface the inherent ductility of the matrix is translated into durability in service. These are the maxims upon which these materials have been developed. I believe that, as the field evolves, these principles will continue to be valid. In the long term we can only afford to use our material resources to their greatest potential.

Appendix 1   Description of laminates

Above the microstructure there is a level of structure in composite materials that is under the control of the product designer. This is the orientation of layers of fibres within a structure. Where the optimum of continuous collimated fibres is employed, the orientation of each layer of a laminated structure is readily defined with respect to a control axis. To identify the construction of such a laminate, we indicate the ply orientation with respect to the control axis and the number of such laminae. Thus 0_4 indicates that four layers of preimpregnated tape are laminated together, with their axis parallel to the control axis (Figure A1.1). By contrast 90_4 implies four layers with their axis transverse to the control axis (Figure A1.2) A set of brackets ( ) indicates a repeated sequence, while a numerical suffix indicates the number of times that the sequence is repeated. The suffix S indicates that the lay up is then symmetrical about the central axis. $(0,90)_{2S}$ could thus be rewritten: 0,90,0,90,90,0,90,0 (Figure A1.3).

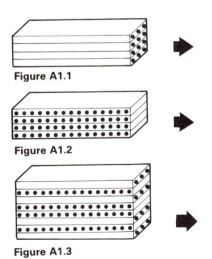

Figure A1.1

Figure A1.2

Figure A1.3

Symmetry about a central axis is of importance in the construction of laminates from highly anisotropic products such as prepreg tape, otherwise thermal stresses may cause distortion.

A so-called quasi-isotropic laminate is a common form that has approximately uniform properties in all directions in the plane of the sheet. Popular forms of such laminates include $(+45,90,-45,0)_{NS}$ and $(0,90,+45,-45)_{NS}$, where N is any number.

Continuous collimated fibre may also be woven into a fabric, and many studies are based on direct impregnation of a woven cloth. For convenience we identify a woven cloth as 0/90 or ±45. Thus $(0/90, ±45)_S$ and $(0,90,+45,-45)_S$ are approximately equivalent quasi-isotropic lay ups in woven and simple laminates respectively. Woven broadgoods may also be prepared from impregnated tapes.

All composites based on continuous collimated fibre possess strong, but definable, anistropy. In the description of basic mechanical properties we shall define the longitudinal properties (along the fibre axis, Figure A1.4), the transverse properties (normal to the fibre axis, Figure A1.5), and a measurement at 45° to the fibre axis based on a symmetrical cross-ply laminate $(+45,-45)_S$, which is a strong indication of shearing within the composite (Figure A1.6).

Figure A1.4   **Figure A1.5**   **Figure A1.6**

Further, we need an indication of shearing within a laminate along the fibre axis (Figure A1.7), and transverse to the fibre axis (Figure A1.8).

Figure A1.7   **Figure A1.8**

Typical stiffnesses for each of these deformations in a carbon fibre reinforced theroplastic composition are shown in Table A1.1.

Table A1.1

Laminate	Tensile modulus @ 23°C GN/m²	Deformation
0_{16} (axial)	135	Figure A1.4
$(+45, -45)_{4S}$	16	Figure A1.6
90_{16} (transverse)	10	Figure A1.5
	Shear modulus @ 23°C GN/m²	
Axial	5.1	Figure A1.7
Transverse	3.2	Figure A1.8

As a first approximation the stiffness of a quasi-isotropic laminate is one third that of the axial stiffness of a uniaxial laminate.

Appendix 2   The importance of shear properties in the matrix

The two primary mechanical duties of the matrix phase are to transfer tensile loading from one fibre to its neighbours and to support the fibres when they are loaded in compression so that they do not buckle.

Consider a composite material where, as a result of some damage, an individual fibre is broken. If there is no matrix, then that fibre would cease to carry load over any of its length with a consequent reduction in stiffness of the composite. However, if a matrix is present connecting that fibre to its neighbours, then the load can be temporarily redistributed to allow that reinforcing fibre to continue to play a valuable role (Figure A2.1).

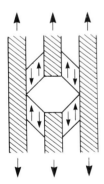

Figure A2.1

This redistribution clearly depends upon the shear properties of the matrix. So effective is the redistribution of stress from a broken fibre to its neighbours that an array of discontinuous aligned fibres, which would readily pull apart in the absence of a matrix, can, through the local redistribution of stress at a fibre ends, have a very similar tensile stiffness and strength to a composite of truly continuous fibres.

The reinforcing fibres are thin. This means that, even though the fibres are stiff, they can be easily bent. Under compressive loading this flexibility would allow an unsupported fibre to buckle, leading to collapse. This buckling is resisted by the matrix resin, which acts to support the fibres and allows them to cooperate to resist the compressive load (Figure A2.2).

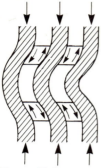

Figure A2.2

This resistance to fibre buckling is provided by the shear properties of the resin and is so effective that a fibrous structure which, without the matrix present, would collapse under its own weight can support an aircraft in flight.

Appendix 3   Mixtures of rods and coils

If the molecules of a molecular composite are of comparable diameter, as is the case with those systems evaluated to date, the spacing of the reinforcing rod molecules must allow penetration by the random coil matrix chains (Figure A3.1).

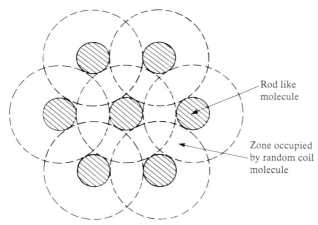

Rod like molecule

Zone occupied by random coil molecule

Figure A3.3

In an ideal system, where the rods are hexagonally packed with the coils penetrating around them, the maximum possible volume fraction of reinforcing rods is about 40%. There is then the problem of threading the random coils molecule through the structure. If a truly random path is followed, the accessible passages soon become blocked, leaving large areas of voided space: many happy hours can be spent with knitting needles and pyjama cord to satisfy oneself of this difficulty.

A possible solution to the problem may exist in the synthesis of giant rod like molecules of diameter about ten times that of the random coil matrix phase: indeed if such molecules were themselves permeable to the random coils and compatible with them, molecular composites could demonstrate their full potential. The chemistry challenge is a significant one.

From the standpoint of small diameter polymers, the use of ternary solutions provides the best prospect of obtaining the optimum single phase consisting of a homogeneous dispersion of rods. Statistical thermodynamics of ternary systems comprising a solvent, rodlike solute and a random coil chain were predicted by Flory[1]. It has been shown that:

1 Aggregates of rigid molecules reject the coiled molecules with a selectivity approaching that of a pure crystal.
2 For a rigid rod molecule with an axial ratio high enough to achieve reinforcement, a high degree of selectivity occurs even at low concentrations.

The theory also provides guidelines for overcoming this problem. Of particular significance is the prediction that, to obtain a homogeneous dispersion, a high degree of molecule orientation of the rod like molecule within the random coil matrix will be required. This may be achieved by recognizing the liquid crystalline properties of the rod like molecules. If the composites are processed from polymer solutions at or near the concentration necessary to form anisotropic domains, then a high dispersion of the rod like molecules can be obtained, and, because the viscosity of the solution is at a maximum, the rod like polymer can be readily oriented by an external shear stress.

Reference

A3–1 P. J. FLORY, 'Statistical Thermodynamics of Mixtures of Rodlike Particles: 5. Mixtures with Random Coils', *Macromolecules,* **11**, pp. 1138–1141 (1978).

Appendix 4 The description of molecular weight and its significance

Molecular weight is usually described in terms of two paramters:

$\overline{M}_N$ = number average molecular weight
$\overline{M}_W$ = weight average molecular weight

The number average defines the total molecular mass divided by the number of molecules present: half the molecules will be longer and half smaller than this size. The weight average molecular weight defines the molecular size where half the total molecular mass is present in larger molecules and half as smaller molecules. Measurements such as gel permeation chromatography describe the number of chains present in a system as a function of their molecular mass, and a typical curve (Figure A4.1) shows a log/normal distribution where $\overline{M}_N$ emphasizes the smaller chains and $\overline{M}_W$ the longer. The ratio $\overline{M}_W/\overline{M}_N$ is a useful characterization of the breadth of the molecular weight distribution.

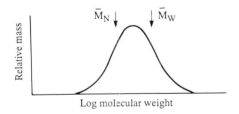

Figure A4.1

If the molecular weight of a polymer is too low, the chains will not entangle sufficiently to achieve a high level of toughness. For a system to be useful all the chains must entangle, and therefore it is the shortest chains in particular that must be above a certain length, making $\overline{M}_N$ the more critical factor for toughness. By contrast, it is the presence of a very few very long chains entangling together that provides the network which must be broken down to achieve flow in the molten state, so that the melt viscosity is largely determined by $\overline{M}_W$. For composite matrices we, in general, require a high toughness and a low viscosity, and to meet these requirements, a narrow molecular weight distribution is preferred.

Appendix 5 Estimation of effective chain diameter

From the molecular model of King and colleagues[1] it is possible to deduce the effective cross sectional area occupied by each molecule when it is packed into a crystal. King shows a unit cell of orthorhombic geometry, with the lattice constants a = 0.783 nm, b = 0.594 nm and c = 0.986 nm (Figure A5.1). where c is along the

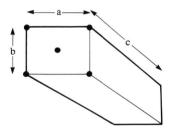

Figure A5.1

length of the chain and contains two phenylene rings (two thirds of the length of the molecular repeat unit). This unit cell contains one full cross section of a molecule and four quarter cross sections, making a total of two full cross sections. The product $\frac{1}{2}$ab is therefore the effective cross-sectional area of the chain in the crystal. For an amorphous chain we must consider the change in density between the crystalline state and the amorphous state:

at 23°C crystalline density 1,400 kg/m³
amorphous density 1,264 kg/m³

The c axis dimension is a virtual invariant. All the density change between crystalline and amorphous phase occurs in the a,b plane. The cross-sectional area of the amorphous chain is then:

$$\frac{1,400}{1,264} \times \tfrac{1}{2}ab = 0.2576 \ (nm)^2$$

Considering the chain as of circular cross section the effective diameter (D) is given by:

$$\frac{\pi D^2}{4} = 0.2576 \ (\text{nm})^2$$

whence $D = 0.57$ nm

References

A5–1 M. A. KING, D. J. BLUNDELL, J. HOWARD, E. A. COLBOURN AND J. KENDRICK, 'Modelling Studies of Crystalline PEEK', *Molecular Simulation*, **4**, pp. 3–13 (1989).

Appendix 6   Radius of gyration of a molecule

The radius of gyration of a polymer chain can be assumed to be equivalent to the mean square end to end distance of a freely rotating chain (r) in an unperturbed state[1].

$$r^2 = m \, l^2 \frac{(1 + \cos \theta)}{(1 - \cos \theta)}$$

where m is the number of rotatable bonds
        l is the bond length
and θ is the supplement to the valence bond angle

In PEEK there are three rotatable bonds in each molecular repeat unit, two ether links and one ketone, whence m is three times the degree of polymerization. The length of the bond is determined by the midpoint spacing of the phenylene units, or one half the c axis length of the crystal unit cell described by King[2]. For PEEK this length is 0.493 nm. King and his colleagues also define the valence bond angle θ as 125°.

For a polymer chain of molecular weight 30,000 the degree of polymerization is 104, whence:

$$m = 312$$
$$l = 0.493$$
$$\theta = 55°, \cos \theta = 0.57$$

thus $r^2 = 312 \times 0.493^2 \dfrac{(1 + 0.57)}{(1 - 0.57)}$ (nm)^2

        $= 277$ (nm)^2

and $r = 16.7$ nm

In a good solvent the valence bond angle will be increased and the radius of gyration swollen accordingly.

References

A6–1 J. BRANDRUP AND E. H. IMMERGUT, *Polymer Handbook,* 2nd edition, J. Wiley (1975).
A6–2 M. A. KING, D. J. BLUNDELL, J. HOWARD, E. A. COLBOURN AND J. KENDRICK, 'Modelling Studies of Crystalline PEEK', *Molecular Simulation,* **4**, pp. 3–13 (1989).

Appendix 7   Some issues in solution impregnation

The use of polymer solutions to wet out the fibre bed has obvious advantages in the ability to achieve a low viscosity state wherein surface tension forces can be relied upon to displace the air at the interface. Usually polymer solutions contain about 10% by volume of polymer and 90% by volume of solvent. The high proportion of solvent further eases wetting of the fibre: to achieve 60% by volume of reinforcement in the final product, the ratios of fibre to polymer solution at impregnation are 60:400, i.e. 13% fibre instead of 60% fibre in the case of simple melt impregnation. To achieve good fibre wetting with solutions in comparison with melts, we therefore have a task that is easier because of the hundred times lower viscosity and the five times lower fibre fraction that is being attempted. The corollary of this easier impregnation is that it is actually more difficult to achieve a controlled fibre loading, since the low viscosity solution readily drains from the fibre surface, leaving a typical volume fraction of fibre in the final product of around 80%, plus, inevitably, some voids. This problem can be offset by adding extra resin to the surface, but this leads to superficially poor resin–fibre distribution in the product.

The second problem area is the elimination of the unwanted solvent from the product. This is achieved by heat and vacuum. The outside surface of the prepreg dries first. This dried surface presents a barrier to further volatilization of the solvent from the inside, which must permeate through the skin. As the solvent evaporates through the skin, it is perhaps natural that the polymer molecules coalesce towards the surface of the prepreg rather than on to the fibres. At some time the interior of the prepreg contains polymer in solution and fibres surrounded by a stiff, and relatively impermeable, skin of polymer that contains a few fibres. At

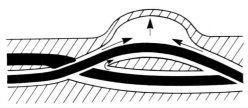

Figure A7.1 Egress of solvent along a fibre interface

this point the remaining solvent may take the easier path of egress by moving to the fibre surface and along that surface until the fibre comes as close to the outside surface as its natural twist will allow, from which point it can burst through and escape (Figure A7.1).

Thus the stage of extracting solvent from a prepreg can result in a qualitative change in fibre wetting and distribution (Figure A7.2).

Figure A7.2 (a) Solution impregnated; (b) solvent extracted

Appendix 8 Thermophysical properties of carbon fibre reinforced polyetheretherketone containing 61% by volume of high strength carbon fibre

Blundell and Willmouth[1] have provided detailed data for the density, transverse thermal conductivity and enthalpy of carbon fibre reinforced PEEK composite over the temperature range 10 to 450°C; Grove and Short[2] have made direct measurement of the coefficient of thermal diffusion, both along and across the fibre direction; and Barnes[3] has extensively characterized the anisotropic thermal expansion. These data, together with other results obtained at ICI Wilton Materials Research Centre[4], are collected in Table A8.1.

Table A8.1

Temperature (°C)	Coefficient of thermal expansion $\times 10^{-6}$/°C		Density kg/m³	Specific heat kJ/kg°C	Coefficient of thermal diffusion 10^{-6} m²/s		Thermal conductivity W/m°C	
	Transverse	Axial			Transverse	Axial	Transverse	Axial
−200	16.4	0.60						
−150	21.6	0.20						
−100	24.6	0.05						
−50	26.4	0.05						
0	28.2	0.15	1,601	0.80	0.33	2.7	0.42	3.5
50	29.6	0.30	1,598	0.93	0.35	3.0	0.52	4.6
100	31.6	0.50	1,593	1.04	0.36	3.1	0.60	5.1
150	36.9	0.20	1,586	1.26	0.36	3.0	0.70	5.9
200	(73.0)	(0.00)	1,575	1.30	0.34	2.9	0.70	5.9
250	(77.0)	(0.00)	1,563	1.40	0.32	2.8	0.70	6.1
300	(84.0)	(0.00)	1,551	1.55	0.31	2.8	0.75	6.7
350	(88.0)	(0.00)	1,537	1.65	0.27	2.7	0.68	6.8
400	(82.0)	(0.00)	1,524	1.70	0.25	2.7	0.65	7.0

( ) Estimated values.

References

A8–1 D. J. BLUNDELL AND F. M. WILLMOUTH, 'Crystalline Morphology of the Matrix of PEEK – Carbon Fibre Aromatic Polymer Composites', *SAMPE Quarterly*, **17**, 2, pp. 50–57 (1986).
A8–2 S. GROVE AND D. SHORT, *Heat Transfer in APC Fabrication*, Plymouth Polytechnic (1984).
A8–3 J. A. BARNES, I. J. SIMMS, G. J. FARROW, D. JACKSON, G. WOSTENHOLM AND B. YATES, 'Thermal Expansion Behaviour of Thermoplastic Composite Materials', *J. Thermoplastic Composite Materials*, **3**, pp. 66–80 (1990).
A8–4 P. H. WILLCOCKS, private communication (1990).

Appendix 9 Flow mechanisms in discontinuous aligned product forms and in crimped fibre systems

Two methods have been proposed for inducing some element of stretchability into continuous fibre composites: long discontinuous fibres and crimped fibres.

The option of long discontinuous fibres[1] (Figure A9.1), permits the individual fibres to slip past one another to allow stretch.

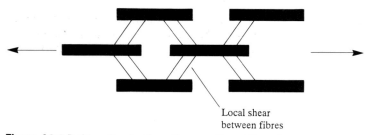

Local shear
between fibres

Figure A9.1 Deformation in discontinuous aligned fibres

At the local level what is actually happening is axial intraply shearing. Pipes and his colleagues[2,3] have extensively analysed the deformation of such structures on the assumption of individual fibre movement, and demonstrate a dependence of viscosity on the square of the fibre length/diameter ratio. Assuming fibre lengths of the order 100 mm and diameters of the order 10 μm, their calculations suggest extensional viscosities in excess of 10^{10} Ns/m^2. Very much lower apparent viscosities are to be expected if the fibre array is presumed to deform inhomogeneously, so that bundles of fibres move co-operatively, thereby dramatically reducing the effective length/diameter ratio (Figure A9.2).

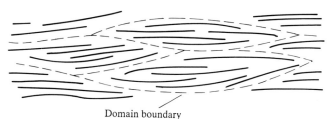

Domain boundary

Figure A9.2 Fibre movement in bundle domains

Such domain movement seems likely to be both statistically and energetically preferred. A consequence of such domain movement is that, locally, the material will thin nonhomogeneously, causing a rise in the local stress and a tendency to rupture if large strains are imposed.

When crimped fibres are used (Figure A9.3), the local deformation is again axial intraply shearing.

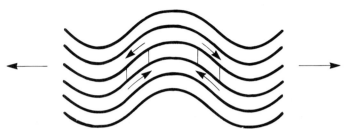

Figure A9.3 Deformation in crimped fibre tapes

In the flow of such materials the fibres are changing from a crimped, easily stretched, state to a fully aligned, unstretchable one, so that resistance to deformation progressively increases. This leads to an automatically self stabilizing deformation process.

References

A9–1 R. K. OKINE, 'Analysis of Forming Parts from Advanced Thermoplastic Sheet Materials', *SAMPE Journal*, **25**, 3, pp. 9–19 (1989).

A9–2 R. B. PIPES, J. W. S. HEARLE, A. J. BEAUSSART AND R. K. OKINE, *Influence of Fiber Length on the Viscous Flow of an Oriented Fiber Assembly*, Center for Composite Materials, University of Delaware (1990).

A9–3 R. B. PIPES, J. W. S. HEARLE, A. J. BEAUSSART AND R. K. OKINE, *A Constituitive Relation for the Viscous Flow of an Oriented Fiber Assembly*, Center for Composite Materials, University of Delaware (1990).

Appendix 10 A comparison of axial and transverse intraply shear viscosities for thermoplastic composites

Isolating the axial and transverse intraply shearing modes (Figure A10.1), for continuous fibres suspended in a viscous medium requires the development of special rheometric techniques. A number of authors have now investigated this problem and a consensus view is emerging (Table A10.1).

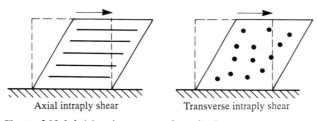

Figure A10.1 Axial and transverse intraply shear

Table A10.1 Intraply shear viscosity (Ns/m²)

Source		Axial mode	Transverse mode	Ratio
Cogswell[1]	CF/PEEK steady flow at 380°C	~6,000	~3,500	1.7
Scobbo[2]	CF/PEEK complex viscosity at 100 rad/sec at 380°C	4,000	3,200	1.3
Cathey[3]	CF/PEEK dynamic viscosity			About 1.0
Groves[4,5]	CF/PEEK dynamic viscosity at shear rate 1 s⁻¹ at 380°C	7,400	6,100	1.2
	Glass/propylene dynamic viscosity at shear rate 1 s⁻¹ at 230°C	860	580	1.5

The large differences in the viscosity of CF/PEEK observed by Scobbo and Nakajima[6] and Groves[7] reflect that they chose to measure the dynamic response at very different angular velocities. Both authors note the dependence of viscosity on angular velocity. Given that dependence, the results are not inconsistent.

The mean value of the ratio of these six independent studies is 1.3 with a coefficient of variation of 0.3.

References

A10–1 F. N. COGSWELL, 'The Processing Science of Thermoplastic Structural Composites', *International Polymer Processing*, **1**, 4, pp. 157–165 (1987).

A10–2 J. J. SCOBBO AND N. NAKAJIMA, 'Dynamic Mechanical Analysis of Molten Thermoplastic/ Continuous Graphite Fiber Composites in Simple Shear Deformation', 21st International SAMPE Technical Conference, pp. 730–743 (1989).

A10–3 C. CATHEY, University of Queensland, private communication (1989).

A10–4 D. J. GROVES, D. M. STOCKS AND A. M. BELLAMY, 'Isotropic and Anistropic Shear Flow in Continuous Fibre Thermoplastic Composites', British Society of Rheology, Edinburgh (1990).

A10–5 F. N. COGSWELL AND D. J. GROVES, 'The Melt Rheology of Continuous Fibre Reinforced Structural Composites Materials', *Proc. Xth International Congress in Rheology*, Sydney (1988).

A10–6 See A10–2.

A10–7 See A10–4.

Appendix 11 Transverse squeezing flow

An analysis of transverse squeezing flow between two plates (Figure A11.1) can be made on the assumption that the resin adheres to the plate surface.

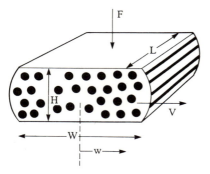

Figure A11.1 Schematic of squeezing flow

F = Applied force
W = Width, across fibre direction
L = Length, along fibre (a constant)
H = Thickness
t = Time
w = Position across the width, w = 0 at centre line, $w = \dfrac{W}{2}$ at free edge

V = Transverse velocity at the edge = $\dfrac{1}{2}\dfrac{dW}{dt}$

There is a pressure gradient P(w) across the width, such that P(w) is zero at W = W/2 and P(w) is a maximum at w = 0.

Consider the flow across an element δw at w

then

$$F = 2 \int_{W/2}^{0} P(w)\, L\, d\, w \qquad \text{(i)}$$

If σ(w) is the shear stress at the surface

then

$$\sigma(w) = \frac{H}{2} \frac{dP}{dw} \qquad \text{(ii)}$$

If the volume flow rate through the element is $Q(w)$ and $\dot{\gamma}(w)$ is the shear rate at the surface, then

$$\dot{\gamma}(w) = \frac{6Q(w)}{LH^2} \qquad \text{(iii)}$$

But

$$Q(w) = \frac{w}{W/2} \; Q(W/2) = \frac{2w}{W} \; LHV \qquad \text{(iv)}$$

From (iii) and (iv)

$$\dot{\gamma}(w) = \frac{12wV}{WH} \qquad \text{(v)}$$

If η_T is the transverse intraply shear viscosity, then

$$\eta_T = \sigma(w)/\dot{\gamma}(w) \qquad \text{(vi)}$$

Combining (ii), (v) and (vi)

$$\frac{dP}{dw} = \eta_T \; \frac{24w}{W} \; \frac{V}{H^2}$$

whence

$$Pw = \int_W^{W/2} \frac{24 \, \eta_T \, V}{WH^2} \; w \, dw$$

$$= \frac{3 \, \eta_T \, VW}{H^2} \left(1 - 4 \frac{w^2}{W^2} \right) \qquad \text{(vii)}$$

Substituting (vii) in (i)

$$F = \frac{2\eta_T \, VW^2 \, L}{H^2} \qquad \text{(viii)}$$

from which

$$\eta_T = \frac{FH^2}{2VW^2 L} \tag{ix}$$

From (v) we can determine the maximum shear rate $\dot{\gamma}_{max}$ as occurring at the free edge

where

$$\dot{\gamma}_{max} = \frac{6V}{H} \tag{x}$$

and since

$$\sigma_{max} = \eta_T \dot{\gamma}_{max}$$

then

$$\sigma_{max} = \frac{3FH}{W^2 L} \tag{xi}$$

Similarly the maximum total shear strain γ_{max} occurs at the free edge, and

$$\gamma_{max} = \int \dot{\gamma}_{max} \, dt \tag{xii}$$

As a first approximation, for small deformation, assuming V and H do not vary significantly, and that

$Vt = D$, the total displacement

then

$$\gamma_{max} = 6D/H \tag{xiii}$$

Appendix 12 Platen pressing of carbon fibre/PEEK laminates

A wide range of moulding cycles can be used when processing CF/PEEK laminates. The following have been found convenient for pressing laminates up to $0.75 \, \text{m}^2$:

1 The prepregs to be pressed should be laid up in the required orientation sequence. A hot soldering iron can be used for tacking the plies together as a unit. For accuracy and handling convenience the unit should be limited to eight plies. Thus several units together may make up the final stack for lamination.

2 The ply stack should be placed between high temper aluminium foils, pretreated with a surface release agent, for example Frekote FRP. To stop material flow at the laminate edges, either a matched mould or picture frame may be used. Because of the low thermal expansion of the composite along the fibre direction, it is preferred that moulds for uniaxial laminates should be slightly oversize (1 mm clearance) along that direction. The mould faces should be able to continue to move closer together as the thickness of the stack is reduced under pressure. Where a frame is used, the foil-faced ply stack should be placed between two mirror-finish stainless steel glazing plates (minimum 1.5 mm thick).

3 The pressing thermal cycle has three distinct freatures: heating, consolidating and cooling. Heating of the stack requires platens with measured surface temperatures between 370 and 390°C. At a nominal contact pressure of one atmosphere the recommended heating time is 1 minute per ply up to a limit of 30 minutes, where a picture frame is used. Consolidation is achieved by applying a pressure (ususally between 10 and 20 atmospheres) to the heated stack for 5 minutes. Cooling should then proceed at the consolidating pressure until the matrix is cold enough to allow release and still maintain consolidation. For PEEK, the moulding should be cooled from a nominal 380°C down to a release temperature of 200°C at a cooling rate of about 40°C/min. This is achieved by transferring the consolidated laminate from a hot (380°C) to a cold (180°C) press for the cooling. The cooling time is about 5 minutes.

In order to achieve good quality mouldings special attention must be paid to platen flatness, sizing of the picture frame, and cooling under sustained pressure.

Appendix 13 Bagging materials for thermoplastic composites

During autoclave consolidation of thermoplastic composites a vacuum bag is required to ensure positive pressure on the part throughout the processing cycle. A typical vacuum bag configuration is illustrated in Figure A13.1. The unconsolidated

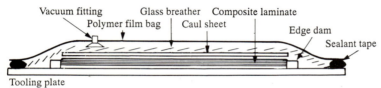

Figure A13.1 Vacuum bag configuration

laminate is placed on the tool plate between edge dams, which have all been treated with high temperature mould release. A caul sheet is then placed on the laminate to isolate it from the breather cloth and ensure a uniform surface finish on the consolidated laminate. Glass breather cloth is placed over the entire laminate stack to facilitate complete evacuation during processing. A thin polymer film is placed over the entire assembly and stuck to the tool plate with a clay like sealant tape. Vacuum is introduced through a high temperature metallic fitting that is attached to the vacuum bag. This entire assembly constitutes the vacuum bag.

The two essential components of the vacuum bag are the polymer film and the sealant tape. The high processing temperatures required for thermoplastic composite consolidation limit the type of materials that can be utilized. In addition, because adhesion is so important, compatibility between the film bagging material and the sealant tape needs to be considered. Listed below in Table A13.1 are different bagging films, compatible sealant tapes, and their temperature limitations.

Table A13.1 Bagging materials for thermoplastic composites

Bagging films	Compatible sealant tapes	Temp. limitations (°C)
Polyimide Films (Upilex, Kapton)	Modified silicone (Airtech A-800) (Schnee-Morehead 5160)	400
Nylon Films (Airtech Wrightlon)	Zinc cromate (Airtech GS 43 MR)	200

Appendix 14　Thermal expansion transverse to the fibre direction

Chapter 4 considered the detailed microstructure of typical composite laminate. It noted the tendency for there to be a resin rich interlayer between plies (Figure A14.1).

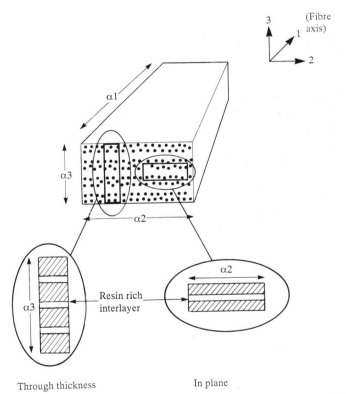

Figure A14.1 Exaggerated sketch of local morphology differences (a) in the plane and (b) through the thickness

The presence of a resin rich layer means that locally the fibres are concentrated to a higher fibre fraction. When considering the case of the in plane thermal expansion, α_2, the composite, at higher volume fraction, and the resin rich interlayer respond in parallel. Those elements are in series in the case of the through thickness expansion. Then if α_{HF} is the transverse thermal expansion coefficient of the high fibre fraction composite (excluding the interlayer), α_M is the thermal expansion coefficient of the matrix in the interlayer, and a is the fractional thickness of the interlayer in the composite,

$$\alpha_3 = a\,\alpha_M + (1 - a)\,\alpha_{HF} \tag{i}$$

$$\frac{1}{\alpha_2} = \frac{a}{\alpha_M} + \frac{(1 - a)}{\alpha_{HF}} \tag{ii}$$

where α_3 and α_{HF} are unknowns. Substituting for α_{HF} from equation (ii) into equation (i) we obtain

$$\alpha_3 = a\,(\alpha_M) + \frac{(1 - a)\,2\,\alpha_2\,\alpha_M}{\alpha_M + a\alpha_2}$$

for a = 0.05 (Chapter 4)
  α_M = 47.0×10^{-6} (Table 6.1)
  α_2 = 28.8×10^{-6} (Table 6.1)
  α_3 = 29.2×10^{-6}

A first estimate of through thickness thermal expansion of angle plied laminates can be made on the assumption that the volume expansion is independent of lay up. This would provide values of 45×10^{-6} and 55×10^{-6} for cross plied and quasi-isotropic laminates respectively. The assumption of constant volume expansion is incorrect, because of the thermally induced volume dilation of the resin resulting from the processing history (Chapter 4). Barnes[1] estimates the through-thickness thermal expansion of cross plied and quasi-isotropic laminates as 38×10^{-6} and 48×10^{-6} respectively. Experimental studies on small samples are consistent with those values.

References

A14–1 J. A. BARNES, private communication (1990).

Appendix 15 Influence of resin-rich interlayers on shear compliance

The same arguments as were used to consider thermal expansion in Appendix 14 can be used to derive the relationship between the in plane, through thickness, and matrix shear compliances.

$$J_{13} = a\,J_M + \frac{(1 - a)^2\,J_{12}\,J_M}{J_M + a\,J_{12}}$$

where: a is the fractional thickness of the resin-rich layer
J_{13} is the through thickness shear compliance = $1/G_{13}$
J_{12} is the in plane shear compliance = $1/G_{12}$
J_M is the shear compliance of the matrix = $1/G_M$

for $G_M = 1.2\,\text{GN/m}^2$ (Chapter 2)
$G_{12} = 5.1\,\text{GN/m}^2$ (Table 6.2)
$ax = 0.05$ (Chapter 4)
$J_{13} = 0.21\,\text{m}^2/\text{GN}$

whence $G_{13} = 4.7\,\text{GN/m}^2$

Appendix 16 Stiffness characteristics of carbon fibre/PEEK composites – a review

This and the following two appendices contain data that describe the mechanical properties of 52% by volume and 61% by volume carbon fibre corresponding to APC-1 and APC-2 respectively. APC-1 was the original development product, and as a result of feedback from various user companies the following changes were incorporated in APC-2: the volume fraction of fibre was increased from 52% to 61%; the processing window, in respect of cooling rate conditions and time in the melt, was broadened to allow slower cooling rates and longer times without over-crystallization or degradation to permit processing by autoclave as well as rapid stamping technology; and the matrix molecular weight was tailored to enhance toughness while maintaining processability, in order to improve the two outstanding properties of damage tolerance and wear resistance further. Both products use high strength carbon fibres: Courtaulds XAS-O in the case of APC-1, and Hercules AS-4 for APC-2. Both products are made by melt impregnation technology and achieve a similar high degree of fibre wetting, so that experience on one system can be translated to another with confidence.

The results reported in these appendices include measurements on preimpregnated materials from a number of independent laboratories, using a range of techniques on samples fabricated by a variety of methods within the manufacturers recommended range. I have not attempted to make judgements about experimental technique, since that is an expert matter.

It should be noted that the results reported here do not include data on the various assembled product forms, such as composites made from commingled fibres. Although the manufacturers data for such products claim that they have equivalent, or even superior, properties to the preimpregnated form, independent assessment (see Chapter 3) indicates that such performance is not usually realized in practice.

Table A16.1 Stiffness of uniaxial laminates at 23°C

Volume fraction fibre (%)	Tensile modulus			Poisson's ratio				Shear modulus		Source (See References)
	E_1 GN/m²	E_2 GN/m²	E_3 GN/m²	v_{12}	v_{13}	v_{21}	v_{23}	G_{12} GN/m²	G_{23} GN/m²	
52	121	10.1		0.37						1
	121	10.1		0.38				4.7		2
	124	8.9		0.38		0.04		4.0		3
								4.2		4
	122	9.2		0.31		0.04		4.5		5
62	135	9.0	9.0	0.34	0.34		0.46	5.2	3.2	6
	134	8.9		0.33	0.30	0.04	0.35	5.1		7
	135	9.2		0.35				4.9		8
	134	8.9		0.31				5.1		9
	152	9.2		0.33				4.4		10
	133	10.3								11
	132	8.9								12
	134	10.0		0.32		0.08		5.3		13
	139	10.4		0.33				4.5		14
	140	9.6						6.2		15

Table A16.2 Stiffness of multi-angle laminates at 23°C

Volume fraction fibre (%)	$(\pm 45)_s$	Tensile modulus (GN/m²) Quasi-isotropic	$(0/90)_s$	Source (See References)
52	12.2	45		16
52	14.5		61	17
52	14.2	47		18
61	18.0	49.5	75	19, 20
61	19.2	47.0	73	21
61		48.6		22
61	19.0			23

Those differences in properties that are not attributable to the difference in fibre content appear to be within the normal scatter of results. Considering that these are independent tests using a variety of test methods in thirteen separate laboratories the high consistency of the results and their agreement with the product data sheet values[24] is worthy of record.

References

A16–1   J. T. HARTNESS, 'An Evaluation of Polyetheretherketone Matrix Composites Fabricated from Unidirectional Prepreg Tape', *SAMPE Journal,* **20**, 5, pp. 26–31 (1984).

A16–2   S. L. DONALDSON, 'Fraction Toughness Testing of Graphite/Epoxy and Graphite/PEEK Composites', *Composites,* **16**, 2, pp. 103–112 (1985).

A16–3   D. P. JONES, D. C. LEACH AND D. R. MOORE, 'Mechanical Properties of Poly Ether Ether Ketone for Engineering Applications', Speciality Polymers Conference, Birmingham, September (1984).

A16–4   S. M. BISHOP, P. T. CURTIS AND G. DOREY, *A Preliminary Assessment of a Carbon Fibre/PEEK Composite,* RAE Technical Report 84061 (1984).

A16–5   F. N. COGSWELL, 'Continuous Fibre Reinforced Thermoplastics', in *Mechanical Properties of Reinforced Thermoplastics,* edited by D. W. Clegg and A. A. Collyer, pp. 83–118, Elsevier (1986).

A16–6   N. M. ZAHLAN, private communication (1990).

A16–7   D. R. CARLILE, D. C. LEACH, D. R. MOORE AND N. ZAHLAN, 'Mechanical Properties of the Carbon Fibre/PEEK Composite APC-2/AS4 for Structural Applications', ASTM Symposium on Advances in Thermoplastic Matrix Composite Materials, Bal Harbour, Florida (1987).

A16–8   A. CERVENKA, 'PEEK/Carbon Fibre Composites: Experimental Evaluation of Unidirectional Laminates and Theoretical Production of their Properties, *Polymer Composites,* **9**, 4, pp. 263–270 (1988).

A16–9   G. JERONIMIDIS AND T. PARKYN, *Residual Stresses in Thermoplastic Matrix Cross-Ply Laminates,* University of Reading (1987).

A16–10 P. J. HINE, B. BREW, R. A. DUCKETT AND I. M. WARD, 'The Fracture Behaviour of Carbon Fibre Reinforced Poly (Ether Etherketone)', *Composites Science and Technology,* **33**, 1, pp. 35–72 (1988).

A16–11 T. NAGUMO, H. NAKAMURA, Y. YOSHIDA AND K. HIRAOKA, 'Evaluation of PEEK Matrix Composites', 32nd International SAMPE Symposium, pp. 486–497 (1987).

A16–12 E. SILVERMAN, R. A. CRIESE AND W. C. FORBES, 'Property Performance of Thermoplastic Composites for Spacecraft Systems', *SAMPE Journal,* **25**, 6, pp. 38–47 (1989).

A16–13 S. GROSSMAN AND M. AMATEAU, 'The Effect of Processing in a Graphite Fibre/Polyetheretherketone Thermoplastic Composite', 33rd International SAMPE Symposium, pp. 681–692 (1988).

A16–14 A. DUTHIE, 'Engineering Substantiation of Fibre Reinforced Thermoplastic Composites for Aerospace Primary Structure', 33rd International SAMPE Symposium, pp. 296–307 (1988).

A16–15 P. VAUTEY, 'Cooling Rate Effects on the Mechanical Properties of a Semi-Crystalline Thermoplastic Composite', *SAMPE Quarterly,* **21**, 2, pp. 23–28 (1990).

A16–16 See A16–1.

A16–17 See A16–4.

A16–18 See A16–5.

A16–19 A. HOROSCHENKOFF, J. BRANDT, J. WARNECKE AND O. S. BRULLER, 'Creep Behaviour of Carbon Fibre Reinforced Polyetheretherketone and Epoxy Resin', SAMPE Conference, Milan, 'New Generation Materials and Processes', pp. 339–349 (1988).

A16–20 A. CERVENKA, 'Mechanical Behaviour of Quasi-isotropic PEEK/Carbon Fibre Laminates: Theory/Experiment Correlation', International Conference 'Advancing with Composites', Milan, 10–12 May (1988).

A16–21 See A16–7.

A16–22 N. ZAHLAN AND F. J. GUILD, 'Computer Aided Design of Thermoplastic Composite Structures', ICCM VII, Beijing (1989).

A16–23 See A16–11.

A16–24 *Aromatic Polymer Composites,* Technical Literature, ICI Fiberite (1986).

Appendix 17 Strengths of carbon fibre/PEEK laminates

Volume fraction high strength CF (%)	Strengths (MN/m²) at 23°C											Source (see References)
	Tension					Compression*				Shear†		
	0	90	±45	0/90	QI	0	90	0/90	QI	0	QI	
52	1,815	87	295		680							1
			361	678					585			2
	1,830		340		700	950				106	80	3
						1,044						4
										111		5
61						1,113						6
	2,130	80	356	1,100	910	1,100				105		7
							675	630				8
						1,102						9
						1,260						10
							192					11
	2,090	73										12
					749							13
			270	1,120								14
	2,350	83	250			1,226	250			127		15
										108		16
	2,240	76										17
	2,131	64				1,116	253			110		18
	2,370	89								102		19
		91										20
	2,270	87				1,140						21

* Compression strength quoted from IITRI testing. Lower values, about 400 MN/m², are observed from the Boeing Compression After Impact Test at zero impact. The latter is a long specimen and may give premature failure by buckling.

† Shear strength is short beam shear, but the failures were not clearly interlaminar shear.

References

A17–1 J. T. HARTNESS, 'An Evaluation of Polyetheretherketone Matrix Composites Fabricated from Unidirectional Prepreg Tape', *SAMPE Journal*, **20**, 5, pp. 26–31 (1984).

A17–2 S. M. BISHOP, P. T. CURTISS AND G. DOREY, *A Preliminary Assessment of a Carbon Fibre/PEEK Composite*, RAE Technical Report 84061 (1984).

A17–3 F. N. COGSWELL, 'Continuous Fibre Reinforced Thermoplastics', in *Mechanical Properties of Reinforced Thermoplastics*, edited by D. W. Clegg and A. A. collyer, pp. 83–118, Elsevier (1986).

A17–4   R. J. LEE, 'Compression Strength of Aligned Carbon Fibre-Reinforced Thermoplastic Laminates', *Composites,* **18**, pp. 1–12 (1987).

A17–5   G. KEMPE AND H. KRAUSS, 'Processing and Mechanical Properties of Fiber-Reinforced Polyetheretherketone (PEEK)', 14th Congress of the International Council of the Aeronautical Sciences, Toulouse (1984).

A17–6   See A17–4.

A17–7   D. R. CARLILE, D. C. LEACH, D. R. MOORE AND N. ZAHLAN, 'Mechanical Properties of the Carbon Fibre/PEEK Composite APC-2/AS4 for Structural Applications', ASTM Symposium on Advances in Thermoplastic Matrix Composite Materials, Bal Harbour, Florida, (1987).

A17–8   Chapter 4.

A17–9   D. LEESER AND D. C. LEACH, 'Compressive Properties of Thermoplastic Matrix Composites', 34th International SAMPE Symposium, pp. 1464–1473 (1989).

A17–10   R. J. LEE AND A. S. TREVETT, 'Compression Strength of Aligned Carbon Fibre Reinforced Thermoplastics Laminates', Sixth International Conference on Composite Materials (ICCM-6), **1**, pp. 278–287 (1987).

A17–11   A. J. CERVENKA, private communication (1990).

A17–12   A. CERVENKA, 'PEEK/Carbon Fibre Composites: Experimental Evaluation of Unidirectional Laminates and Theoretical Production of their Properties, *Polymer Composites,* **9**, 4, pp. 263–270 (1988).

A17–13   A. CERVENKA, 'Mechanical Behaviour of Quasi-isotropic PEEK/Carbon Fibre Laminates: Theory/Experiment Correlation', International Conference 'Advancing with Composites', Milan (1988).

A17–14   A. CERVENKA, 'Composite Modelling: Polyetheretherketone/Carbon Fibre Composite (APC-2) Experimental Evaluation of Cross Plied Laminates and Theoretical Prediction of their Properties', Rolduc Polymer Meeting-2, Limburg (1987).

A17–15   T. NAKAMURA, Y. YOSHIDA AND K. HIRAOKA, 'Evaluation of PEEK Matrix Composites', 32nd International SAMPE Symposium, pp. 486–497 (1987).

A17–16   See A17–5.

A17–17   S. GROSSMAN AND M. AMATEAU, 'The Effect of Processing in a Graphite Fibre/Polyetheretherketone Thermoplastic Composite', 33rd International SAMPE Symposium, pp. 681–692 (1988).

A17–18   A. DUTHIE, 'Engineering Substantiation of Fibre Reinforced Thermoplastic Composites for Aerospace Primary Structure', 33rd International SAMPE Symposium, pp. 296–307 (1988).

A17–19   M. NARDIN, E. M. ASLOUN, J. SCHULTZ, J. BRAND AND H. RICHTER, 'Investigations on the Interface and Mechanical Properties of Composites based on Non-Impregnated Carbon Fibre Reinforced Thermoplastic Preforms', *Proceedings 11th International European Chapter of SAMPE,* pp. 281–292 (1990).

A17–20   E. SILVERMAN, R. A. CRIESE AND W. C. FORBES, 'Property Performance of Thermoplastic Composites for Spacecraft Systems', *SAMPE Jounal,* **25**, 6, pp. 38–47 (1989).

A17–21   P. VAUTEY, 'Cooling Rate Effects on the Mechanical Properties of a Semi-Crystalline Thermoplastic Composite', *SAMPE Quarterly,* **21**, 1, pp. 23–28 (1990).

Appendix 18 Fracture toughness data on carbon fibre/PEEK composites at 23°C

Product form	Strain energy release rate		Source (see References)
	Mode I G_{IC} (KJ/m^2)	Mode II G_{IIC} (KJ/m^2)	
Uniaxial tape	1.9–3.2		1
52% by volume	1.1	1.1	2, 3
carbon fibre	1.4		4
Uniaxial tape	1.4	1.8	5
61% by volume	1.2	1.8	6
carbon fibre	1.9		7
	3.4		8
	2.4–2.9		9, 10
		1.9	11
	2.0		12
		2.2	13
	2.0	2.4	14
	1.7		15
	2.5	2.4	16
	1.5–2.0		17
	2.5		18
	2.0	2.4	19
	1.6	1.2	20
	1.0–2.2		21
	1.1–1.9		22
		2.0	23
		1.3–2.0	24
	1.5–2.4	2.8	25
	2.2		26
	1.7	1.0–2.2	27
	1.8, 2.4, 2.5		28
	2.9, 3.4, 3.5		29
		1.7	30
Woven fabric	1.3		31
55% by volume	1.5–4.5		32
carbon fibre			
MEAN (SD)	2.1 (0.8)	1.8 (0.5)	

Individual authors report a coefficient of variation of approximately 10% for any individual test. The variation in results reported here reflects differences in test method, in interpretation of the test results, and in the rate of testing. In some cases cracks propagate in an unstable manner: a relatively slow crack growth with high strain energy release and gross ductility in the matrix is followed by spurts of rapid growth at a lower rate of energy dissipation[33]. At lower temperatures or faster crack growth rates the less ductile mode is more evident, but failure is always within the matrix phase not at the fibre/matrix interface.

References

A18–1   D. R. CARLILE AND D. C. LEACH, 'Damage and Notch Sensitivity of Graphite/PEEK Composite', 15th SAMPE Technical Conference Cincinatti (1983).

A18–2   S. L. DONALDSON, 'On Fracture Toughness Testing of Graphite/Epoxy and Graphite/PEEK Composites', *Composites*, **18**, 1 (1987).

A18–3   S. L. DONALDSON, 'Fracture Toughness Testing of Graphite/Epoxy and Graphite/PEEK Composites', *Composites*, **16**, 2, pp. 103–112 (1985).

A18–4   N. SELA AND O. ISHAI, 'Interlaminar Fracture Toughness and Toughening of Laminated Composite Materials: A Review', *Composites*, **20**, 5, pp. 423–435 (1989).

A18–5   *Ibid.*

A18–6   L. BERGLUND AND T. JOHANNESSON, 'Mixed-mode Fracture of PEEK/Carbon Fibre Composites', First European Conference on Composite Materials, Bordeaux (1983).

A18–7   D. C. LEACH AND D. R. MOORE, 'Toughness of Aromatic Polymer Composites', *Composites Science and Technology*, **23**, pp. 131–161 (1985).

A18–8   D. R. CARLILE, D. C. LEACH, D. R. MOORE AND N. ZAHLAN, 'Mechanical Properties of the Carbon Fibre/PEEK Composite APC-2/AS4 for Structural Applications', ASTM Symposium on Advances in Thermoplastic Matrix Composite Materials, Bal Harbour, Florida (1987).

A18–9   D. C. CURTIS, D. C. LEACH AND D. R. TAMBLIN, 'The Delamination Behaviour of Aromatic Polymer Composite APC-2', ASTM Symposium on Toughened Composites, Houston (1985).

A18–10 R. A. CRICK, D. C. LEACH AND D. R. MOORE, 'Interpretation of Toughness in Aromatic Polymer Composites using a Fracture Mechanics Approach', 31st International SAMPE Symposium, pp. 1677–1689 (1986).

A18–11 B. R. TRETHEWEY JR, L. A. CARLSSON, J. W. GILLESPIE JR AND R. B. PIPES, *Mode II Interlaminar Fracture During Static and Fatigue Loading*, CCM86–26, University of Delaware (1986).

A18–12 C. Y. BARLOW AND A. H. WINDLE, *The Measurement of Fracture Energy in Aligned Composites*, University of Cambridge, pp. 1–11 (1986).

A18–13 B. L. SMITH, 'Delamination in Laminated Composite Materials', 18th International SAMPE Technical Conference (1986).

A18–14 A. J. SMILEY AND R. B. PIPES, 'Rate Sensitivity of Interlaminar Fracture Toughness in Composite Materials', *Proc. American Soc. for Composites*, First Technical Conference, Dayton, pp. 434–449 (1986).

A18–15 J. W. GILLESPIE JR, L. A. CARLSSON AND A. J. SMILEY, *Rate Dependent Mode I Interlaminar Crack Growth Mechanisms in Graphite/Epoxy and Graphite/PEEK*, University of Delaware (1987).

A18–16 S. HASHEMI, A. J. KINLOCH AND J. G. WILLIAMS, 'Interlaminar Fracture of Composite Materials', ICCM-6, **3**, pp. 254–264 (1987).

A18–17 P. J. HINE, B. BREW, R. A. DUCKETT AND I. M. WARD, 'Failure Mechanisms in Continuous Fibre Reinforced PEEK', ICCM-6, **3**, pp. 397–404 (1987).

A18–18 P. DAVIES AND F. X. DE CHARENTENAY, 'The Effect of Temperature on the Interlaminar Fracture of Tough Composites', ICCM-6, **3**, pp. 284–294 (1987).

A18–19 T. J. CHAPMAN, A. J. SMILEY AND R. B. PIPES, 'Rate and Temperature Effects on Mode II Interlaminar Fracture Toughness in Composite Materials', ICCM-6, **3**, pp. 295–304 (1987).

A18–20 J. W. GILLESPIE AND L. A. CARLSSON, *Buckling and Growth of Delamination in Thermoset and Thermoplastic Composites*, University of Delaware (1989).

A18–21 A. L. GLESSNER, M. T. TAKEMORI, S. K. GIFFORD AND M. A. VALLANCE, 'Temperature and Rate Dependence of Delamination in Aerospace Composite Laminates', SPE ANTEC '87, pp. 1454–1457 (1987).

A18–22 P. J. HINE, B. BREW, R. A. DUCKETT AND I. M. WARD, 'The Fracture Behaviour of Carbon Fibre Reinforced Poly (Ether Etherketone)', *Composites Science and Technology*, **33**, 1, pp. 35–72 (1988).

A18–23 H. MAIKUMA, J. W. GILLESPIE AND J. M. WHITNEY, 'Analysis and Experimental Characterization of the Center Notch Flexural Test Specimen for Mode II Interlaminar Fracture', submitted to *J. Composite Materials* (1989).

A18–24 H. MAIKUMA, J. W. GILLESPIE AND D. J. WILKINS, 'Mode II Interlaminar Fracture of the Center Notch Flexural Specimen under Impact Loading', submitted to *J. Composite Materials* (1989).

A18–25 P. VAUTEY, 'Cooling Rate Effects on the Mechanical Properties of a Semi-Crystalline Thermoplastic Composite', *SAMPE Quarterly,* **21**, 2, pp. 23–28 (1990).

A18–26 L. A. BERGLUND AND T. R. JOHANNESSON, *Effects of Deformation Rate on the Mode Interlaminar Crack Proppagation in Carbon Fiber/PEEK Composites,* Linkoping Institute of Technology, pp. 131–159 (1987).

A18–27 P. DAVIES, W. J. CANTWELL, P.-Y. JAR, H. RICHARD, D. J. NEVILLE AND H. H. KAUSCH, 'Cooling Rate Effects in Carbon Fibre/PEEK Composites', 3rd ASTM Symposium on Composite Materials, Orlando (1989).

A18–28 See A18–10.

A18–29 See A18–10.

A18–30 T. J. CHAPMAN, 'Temperature and Rate Effects on Mode II Interlaminar Fracture Toughness of Composite Materials', B.Sc Thesis, University of Delaware (1986).

A18–31 J. T. HARTNESS, 'Polyether etherketone Matrix Composites', 14th National SAMPE Technical Conference, **14**, pp. 26–32 (1982).

A18–32 G. DOREY, 'Impact Damage in Composites – Development, Consequences and Prevention', ICCM-6, **3**, pp. 1–26 (1987).

A18–33 See A18–10.

Subject Index

Author Index

Listed with reference numbers.